PROSPECTING AND OPERATING SMALL GOLD PLACERS

Prospecting and Operating Small Gold Placers

BY

WILLIAM F. BOERICKE
Mining Engineer

SECOND EDITION

JOHN WILEY & SONS, INC.

NEW YORK   LONDON   SYDNEY

Printed in the U. S. A.

To
ARTHUR THACHER
Mining Engineer

whose counsel and leadership
have inspired all of us who
were privileged to work with
him

CONTENTS

CHAPTER   PAGE

   INTRODUCTION i

I HOW PLACER DEPOSITS HAVE BEEN FORMED...... 1

II EXAMINING PLACER GROUND. HINTS ON PROS-
   PECTING 12

III PANNING. PROPERTIES OF GOLD AND BLACK SANDS.
   ESTIMATING THE VALUE OF COLORS........... 25

IV THE ROCKER: ITS CONSTRUCTION AND OPERATION 36

V GENERAL INFORMATION ON SLUICING............ 49

VI SLUICE BOX CONSTRUCTION. THE LONG-TOM. TYPES
   OF RIFFLES 57

VII METHODS OF WORKING PLACER GROUND.......... 69

VIII CLEANING-UP. RECOVERY OF FINE GOLD FROM
   BLACK SANDS. UNDERCURRENTS AND SIMILAR
   DEVICES 82

IX USE OF MERCURY IN PLACER MINING. AMALGAMA-
   TION. RETORTING 94

X DRY PLACERS 101

XI PLACER MINING MACHINES.................... 109

XII LOCATION OF PLACER MINING CLAIMS. LEASING.
   SALE OF PLACER GOLD. SELLING A PLACER PROP-
   ERTY 121

   BIBLIOGRAPHY 134

   APPENDIX 137

   INDEX 141

INTRODUCTION

"The winning of gold from alluvial material," said C. W. Purington in his classic monograph on Placer Mining in Alaska, "is a business difficult both to learn and conduct successfully. The careful miner will give as much attention to one part of his business as to another, irrespective of the scale on which it is conducted."

No better counsel can be given to the thousands of would-be placer miners whose imaginations have been stimulated by loosely written articles that have appeared in the daily press since 1930, wherein it is too often implied that the West is full of placer ground where good wages, possibly a fortune, can be made by men of no experience who are willing to go out and prospect.

That some placer areas of promise await discovery is undoubtedly true. That the inexperienced man, ignorant of prospecting and operation, can find them and open them up profitably is certainly over-optimistic, particularly if he has little or no money to put into equipment for economical operation and is unfamiliar with the best methods to attack them.

Both the Bureau of Mines and various state mining bureaus have done commendable work in publishing monographs and bulletins designed to aid the inexperienced man in prospecting for gold placers, as well as indicating where the most hopeful areas are likely to be found. These publications are listed in the Appendix.

This handbook has been written to assist the man without technical education in prospecting such areas intelligently, and to suggest means for equipping and operating the placers at the least cost to show a profit on the outlay of labor and money. Some of the methods described demand little more than native ingenuity, a little cash, and a strong back. Others will require the investment of a modest amount of money for equipment. No space in this book has been devoted to description of methods that involve substantial investments of capital, however important they may be in contributing to national output of gold. For this reason, dredging and hydraulicking, the largest factors in the mining of alluvial deposits, have received small mention in these pages, as the amount of capital and engineering experience required for them put them out of reach of the small operator.

The methods and equipment described in this book are recommended as having proved their merit in scores of camps; they have unquestionably been responsible for an important proportion of the world's gold. In obtaining the material for the text I have consulted the publications of the Geological Survey and the Bureau of Mines, as well as those of the states where placer mining is important. The technical press and professional papers of the American Institute of Mining and Metallurgical Engineers have yielded much valuable material. Constructive criticism of the text has been received from several experienced placer engineers, in addition to correspondence from others in the field, for all of which I am indebted.

If this book shall contribute to making the work of the courageous men who go out into the hills in search of gold a little easier and more productive, and keep them from making costly errors in developing and operating their ground, my purpose will have been fully attained.

WILLIAM F. BOERICKE

New York
March, 1933

FOREWORD TO SECOND EDITION

The change in the price of gold from the old statutory figure of $20.67 per ounce to $35 makes necessary the revision of many statements in the first edition as to the value of gravel and gold particles having a stated content in units of weight, whether expressed in the English or metric system. This change has been made in all instances, as I believe there will be less confusion in accepting the present price of $35 per ounce, even though not yet stabilized by world agreement, than to inconvenience the reader by making it necessary for him to multiply all figures by 1.69 to get at the present value.

The chapter on Placer Mining Machines has been entirely re-written with much new material added, and descriptions of a few new machines that have demonstrated their merit in the field are included. A host of new devices for gold saving of more or less merit have been patented and offered for sale; most of them have failed to justify the hopes of their inventors when subjected to the hard reality of practical operation in the field.

My thanks are due to the many men who have taken the trouble to write me of results obtained in their work, and to offer valuable suggestions for this second edition.

PLACER MINING

CHAPTER I.

HOW PLACER DEPOSITS HAVE BEEN FORMED.

The man who goes in search of placer gold need not have a profound knowledge of geology, nor qualify as an expert mineralogist. But if he is to avoid much needless prospecting and waste of time and money, he should have some clear idea of how placers originate, the different types that he is likely to encounter, and their comparative value as a source of gold.

Geological theory, proved true by the history of thousands of placer deposits, and the experience of legions of miners, asserts that placers are always of secondary, not primary origin. They are unconsolidated accumulations of rock and mineral fragments, with boulders, clay, sand, and gravel, in which the gold forms but an infinitesimal percentage. Nature did not create a placer as she did a vein or lode. In the beginning the gold existed in the vein itself. It may have been present in a single vein that outcropped boldly on a hillside, or there may have been a whole series or network of small veins, which individually had only small amounts of gold. Through countless ages the surrounding country rock and vein material were broken down and eroded away by rain, frost, and chemical action. The gold, being chemically resistant and heavy, was left behind. This was the first step in the

1

formation of a placer—the freeing of the gold from the surrounding country rock by natural agencies. This process is going on today, but at so inconceivably slow a rate as to be unnoticed by the human eye.

Saprolite deposits

If the formation of a placer stopped at this point, as it sometimes does, there would be little concentration of the gold. The richest portion of the deposit would already be at the surface, and the values would get progressively leaner as we went deeper. Such occurrences of gold are common in the

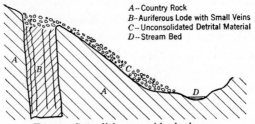

A -- Country Rock
B -- Auriferous Lode with Small Veins
C -- Unconsolidated Detrital Material
D -- Stream Bed

Fig. 1.—Saprolitic or residual placer.

Appalachian regions of the southern states, and are called saprolitic or residual deposits. They occur in upland areas. A typical occurrence of residual placers in eastern Nicaragua is illustrated in Fig. 1.* Here the country rock containing small veins of gold and heavily mineralized with pyrite has been completely decomposed from the surface to a depth of 30 feet, with a gradual transition below the 30-foot layer of complete decomposition into unaltered rock 200 to 300 feet below. The rainfall, extremely heavy, about 140 inches, washes the decomposed rock down the long sloping ridges which rise

* Correspondence, C. C. Semple, E. M.

only a few hundred feet above the surrounding flat country, with the result that the top of the hill is nearly bare, and the detrital material collects at or near the base in a zone much wider than that of the gold-bearing rock or vein. This layer, a sort of mushroomed migration of the outcrop, is locally called a "manta" or blanket. During the slow migration of the material down the sides of the hill there was not enough water for any sorting action to take place, so the gold is found pretty well distributed throughout the manta—in which respect it differs markedly from a true placer deposit. The particles of gold are rough and jagged with no evidence of being waterworn, indicating nearness to the original lode. Unlike the saprolitic deposits of the southern states where much refractory clay is mixed with the gravel, this material is friable and can be broken up without much difficulty by water.

Such conditions are generally rare with this type of placer deposit, and because of the small amount of concentration, the limited depth, and the difficult working conditions, it generally proves disappointing when more than superficial prospecting is employed. Where it occurs in situ, the concentrated material rests on the outcrop; and if the metallic particles are closely packed, they form a protecting blanket which retards or prevents entirely any further growth of the enriched layer.

Conditions under which a true placer is formed

In the formation of a true placer deposit, the second step involves the transportation of the eroded material by running water. If the original vein or veins were on a hillside draining down to a swiftly running stream, a powerful factor is added.

Boulders not yet freed from the gold are pounded up and pulverized, but the gold, being soft and ductile, is merely rounded and flattened into flakes and scales and left behind.

Running water performs a third important service in placer formation. It sorts the material with which it is laden into different strata, according to the specific gravity of such material. As the velocity of the stream diminishes when the grade lessens or a bend is encountered, some of the heavier material drops to the stream bottom. Through natural classification, the heavier material tends to work its way downward through the gravel bed, until it finds a final resting place on bedrock, or on an impervious clay stratum that holds it. The bottom layers of stream gravel are consequently enriched with heavy metallic particles, the coarse gold seeking the bedrock while the fines are higher up. But even on the bedrock, if it is smooth, the gold will not remain permanently, but will work slowly downstream, until it lodges against a ledge or some other natural riffle. At the Breckenridge, Colorado, placers, concentration of the gold was frequently found behind the shales and slates that were standing on edge in the stream beds. Where these ledges were tilted over in the direction of the water flow, the gold would occur on the under side, where there was an eddying action.

Reviewing then the steps in the formation of a true placer deposit, there is first erosion of gold-bearing country rock on a hill or mountain side; second, separation and transportation of the material by rains to a running stream; third, sorting, classification, and concentration of the material in water; and fourth, deposition of the gold in the stream bed owing to some slackening or interference with the water flow.

Bench placers

This is the typical way that a stream placer has been formed. There are other types of deposit to consider as well. Placers may be found considerably removed from the present

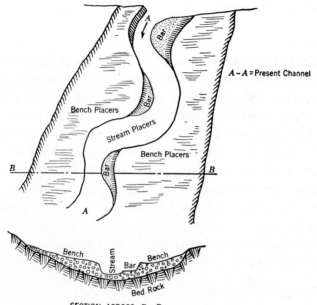

A—A = Present Channel

SECTION ACROSS *B - B*
FIG. 2.—Bar, bench, and stream placers.

course of the stream. They may be found many feet above the present water level, representing the gravel deposits in a former channel which has been long since left high and dry, the stream having changed its course or cut a new channel through the old bed. (See Fig. 2.) These latter, known as bench placers, have been a prolific source of gold. Bench placers may be 50 to 300 feet above the present stream placers and bear little relation to the present course of the stream, yet have all the characteristics of modern stream placers. They may be many feet in thickness, with bedrock above that of the stream. Hillside placers are intermediate between bench and stream placers.

Bar placers

River bar placers are formed on the larger streams during periods of high water, and left exposed when the water subsides. They are seldom permanent and tend to shift downstream. The gold is usually found on the upstream end; the particles are finely divided and difficult to save, but the dissemination is more uniform than in the other types. It is a general rule that the placers found along the lower parts of a stream are not so rich as those in the upper portions, on account of the more thorough mixing of the fine gold through the whole deposit as it moves onward, and its comminution on its travels. However, they serve as a good indication of better ground in the tributary streams.

Other types of placers

Some other types may be mentioned, though they do not fall under the usual head of placer mining. Buried placers are old river channels, of prehistoric times, that have been elevated and subsequently buried under a lava flow. They must be worked by underground mining methods, and their prospecting and development usually constitute a job for the engineer. Dry placers occur in desert areas; their geology and occurrence are discussed in Chapter X. Beach placers occur at various points on the Pacific coast and have been a prolific source of gold at Nome. They are really a re-sorted river deposit, in which the action of the tide has concentrated the gold in the sands into strata that can be worked profitably by hand, rather than large scale methods.

Appearance of gold in stream placers

By far the most important type for the small operator is the stream placer, which is also the most widespread. It may be confidently looked for in those areas where gold-bearing veins are present, though it does not necessarily follow that rich placers are found in close proximity to rich lode mines any more than the occurrence of workable placer deposits indicates that the original source of the gold is near by. Indeed, the conditions favorable for concentration of the gold may be found many miles from where the precious metal was first torn out from the vein walls, and it is known that many placers that contain pay-gravel are the concentration of many small gold veins, none of which would pay to work by itself.

Some indication of the proximity of the placer to the mother lode can be gained by a close examination, under a microscope, of the gold grains. If the particles are rounded and waterworn, they have obviously been carried a long way by the stream and in their travels have had the adhering rock material worn off, leaving the soft gold to be rounded into spherical shape. If the particles are ragged and angular, with slivers of quartz still sticking to the metal, the original vein must be reasonably close by. Flat, paper-thin colors (flake gold) may have been transported hundreds of miles by a swift stream or muddy water.

Favorable places to prospect

Ideal conditions for formation of stream placers occur in valleys leading off the drainage from mountains with auriferous lodes, where there is evidence of considerable erosion, followed by a moderate uplift, and no evidence of subsequent glaciation. The coarsest gold occurs usually at the heads of ravines or gulches, unless the gradient of the stream is too great—which would result in all material being swept clean along the bedrock. Rich deposits are seldom opposite the mouths of tributary streams because the increased volume of water and greater velocity of the current tend to sweep any material onward. Nor is it certain that concentration will be found in potholes below rapids, which at first thought might seem to be ideal collectors. Such potholes in time of floods act as first-class milling machines; any material caught in them is subjected to excessive circulation and grinding, and the gold is liable to be thrown out by the whirling action, to be deposited in quieter waters below. A too rapid erosion with a swiftly running stream presents little chance for sorting and concentration of the gold, hence streams of moderate gradient, about a 30-foot fall to the mile, offer a better chance to the prospector.

Position of the pay-streak

The distribution of gold in the gravels of a placer is extremely irregular; the pay-streak seldom follows a set course, and it may lie in a direction quite different from that of the present stream bed. This follows from the fact that the original course of the stream when the gold was laid down may have been, and probably was, quite different from its present course. There may be two or more pay-streaks in the channel, not necessarily in the deepest part of the gulch, but along the sides. Nearly always there is a sudden peaking of values near a high-grade spot, with a quick drop-off as it is left. In a narrow V-shaped valley, there is a likelihood that the gold may be found across the whole stream bottom; but in a broad, flat-shaped valley, the pay-streak is likely to be narrow, and to be found on one side or the other of the stream.

Although in a true placer the heaviest concentration of gold is invariably at or near the bedrock, the position of the pay-channel may not be in the deepest part. Concentration is usually heaviest on the inner side of the bends of the stream, whether ancient or modern, where the velocity of the water has slackened, and the river bed has widened. The pay-streak is not necessarily continuous, as the points favorable for deposition on the stream bed, such as natural riffles, breaks in the bedrock, bends and bars, will be interrupted. If the bedrock itself is fractured, and at the same time the slope flattens out so that the stream velocity is cut down, conditions are good for concentration, and in such cases it will pay to take up all the decomposed bedrock for treatment.

Suggestions for prospecting

The prospector can be sure of no fixed rule for the location of the pay-streak on the stream bed. It is a case of cut and try. In general, he should test carefully those places where there is indication of the stream widening out, or changing slightly its direction of flow, because obviously the velocity was slowed up at such points. As the movement of the water is always swiftest at the center of the stream and slower along the shores, there is a better chance for deposition along the latter, particularly after high water. As a placer deposit may be shifted many times through natural agencies, the pay-streak may shift accordingly. Uniform dissemination of values is exceptional. Sometimes even the top gravel pays to work. Sand generally offers poorer prospects than gravel; clay, of course, need not be considered, except on the top or bottom, where it may act as a collector of fine gold.

Gulch and creek placers have proved in the past the most productive source of placer gold for the man with limited capital. They are the most easily discovered; they require the minimum amount of money for working; the gold is usually coarser and more easily recovered with the simple methods of working; and there is a greater likelihood of bonanza pockets to serve as a lure.

Flood gold

Occasionally one hears of "flood gold." This is a general term used by engineers to describe an extremely thin film of gold, usually finely divided or flour-like in size, that appears on river bars after high water, and creates great excitement because it is superficially abundant and apparently rich. Unfortunately, there is seldom any concentration, and after the surface has been skimmed off, nothing is left. There may or may not be a true placer below it; its appearance is no guarantee that there is. Economically it is of small moment, and after another flood, it wanders on its errant path. However, in some localities it is quite recurrent, and on certain rivers the "crop" is regularly worked by natives. A somewhat similar occurrence is found in certain saprolitic areas, where after a heavy rain, specks of gold can be picked up on the ground. In Australia this practice of "specking" is quite common. "Moss miners," in some auriferous rivers in the West, at times of low water collect the moss that grows along the river bed which has entangled within it some of the flood gold of high water; they burn this and pan the residue, which sometimes yields them a surprising amount of gold.

Chapter II.

EXAMINING PLACER GROUND. HINTS ON PROSPECTING.

Preliminary investigation of placer ground is necessary even for the smallest-scale operations, to avoid loss of time and money in putting in equipment that may not subsequently be justified. Although a miner with nothing but a shovel and pan will, of course, combine his prospecting with the development of his claim, working out the richer ground as fast as it is uncovered, and will not find it necessary to measure yardage and calculate yield if his pannings show up good pay, it will be a satisfaction to know that he has more than a few hours' or a few days' work ahead of him. For any more ambitious scheme of working, after the presence of gold has been discovered in the gravel, it is necessary to make some approximately accurate estimates of the value of the ground before proceeding farther.

Such evaluation involves some measurement of the yardage of gravel available, its gold content, the percentage that can be recovered, and a consideration of the various economic factors that affect the choice of equipment, such as character of gravel, whether loose, sandy, clayey, or cemented, the size and number of boulders, water supply, grade of bedrock, dumping room, etc. While experience is the best guide in such matters, common sense and keen observation will keep the newcomer from going far wrong.

Test pits

If the placer is of the creek or gulch type, an effort should be made to determine the course of the channel and the depth and position of the bedrock by sinking small test pits through the gravel at points ahead and to the sides so as to delimit it. These pits should, of course, be sunk to the bedrock, and if it is fractured or decomposed, through such soft material down to the solid rock. Care should be taken not to stop at false bedrock, or a hard stratum of clay or hardpan that may lie above it.

The material taken out of such pits should all be panned, or put through the rocker, and the count kept of the number of colors, so that a record will be had not only of every foot of depth, but also of the entire contents of the pit. If this involves too much work, a box of convenient size to hold a measured quantity of gravel can be used, say, 2 ft. by 2½ ft. in cross-section, and 2 feet deep. This will hold 10 cubic feet, about one-third of a cubic yard in place, allowing 20 per cent for the "swell" of the gravel, and if a fair sample of the gravel is taken, a rough estimate of the ground can be calculated.

Hints on sampling

Accurate sampling of placer ground by this means is well-nigh an impossible job. Not the least difficulty is to estimate the amount of boulders in the gravel, which, of course, interfere with taking a fair sample. About the only thing to do is to estimate by eye the number of boulders larger than 3 to 4 inches in the gravel, and use this factor in correcting the value of the ground as shown by the panned sample. For instance, if there seems to be 50 per cent of boulders that cannot be taken in the pan, and a measured cubic yard of the fine gravel has 50 cents in gold, the average value of the ground will be 25 cents. Actually, it will be between 25 and 50 cents, for some of the fine dirt on the boulders will contain gold which will be recovered in sluicing or washing.

Errors in sampling can be lessened by sinking small shafts and putting all the material through the rocker. If the cross-section of the shaft is made 2 ft. 3 in. by 4 ft., then 3 feet of shaft equals 1 cubic yard in place. If the shaft is 9 feet deep to bedrock, the material put through the rocker will represent 3 cubic yards, or 1 square yard of bedrock area, and there is no need at all of measuring it further by box measurements.

The advantage of sinking pits for sampling over any other method, aside from greater accuracy, is that it permits close examination of the character of the gravel at the same time as well. In most cases the overburden of gravel above the pay-streak carries only small values; this point can be definitely determined by sampling, and means may be considered for eliminating the valueless material in operation. If, in prospecting bench placers, or deep channels, the gravel exceeds 20 feet in thickness, shaft sinking becomes expensive, and if water is present, almost impossible without heavy pumping expense. In such cases, recourse must be had to drilling, with drills of the Empire type; but this work is not within the province of the small operator for whom this book is written, and it should have the supervision of an experienced engineer. To drill a sizable piece of placer ground with any degree of thoroughness costs from $5,000 to $10,000.

In most cases, after the pay-channel has been located, it is necessary to work it by an open cut that will not extend all the way across the deposit. If the upper portions have little value, the pay-gravel must have enough gold to carry the expense of all the poor ground that lies above it, in addition to the slope of the sides. An instance is presented in a Colombia placer of a bank 10 to 30 feet in height where the top gravel averaged only 5 to 10 cents per yard from the surface down to the last 2 or 3 feet above bedrock, when the pay-streak, 3 feet in thickness, averaged $1.00.

If there is an open face of the gravel, it can be sampled roughly by cleaning off a 2-foot section vertically, measuring off 1-foot intervals, and taking equal samples straight up and down, which can be panned separately, and the values calculated, assuming a certain number of pans per cubic yard, and counting or weighing colors. More data on this procedure are given in Chapter III.

Naturally in estimating value of ground, full consideration is given to results obtained in neighboring placers where conditions appear to be similar, as well as to any records that may be available from previous attempts to work the ground. In this connection, do not underrate the work of the old-timers. They generally knew their business, and attempts to rework old ground, with the hope that modern methods will show vastly improved results, usually end in disappointment.

Character of gravel

Particular attention should be given to the character of the gravel, and the number and size of boulders. If these are large and abundant, the cost of removing them so as to work the ground may be prohibitive. If much clay is present there is sure to be trouble with disintegrating the gravel, and more water will be needed than usual. Sand layers are usually barren, or nearly so. If the gravel is clean, it is likely that the descent of the gold to bedrock was not hindered, and concentration may be expected. The presence of much black sand may or may not be a good indication of gold. It is usually present in amounts averaging about 35 pounds per yard of gravel, but it may be much more. Excessive black sand tends to choke up the riffles, and makes recovery of fine gold difficult. The physical appearance of the gravel is often suggestive: ancient gravels are abundant with quartz fragments; modern gravels are darker in color.

Fire assays

In very few cases is it advisable or necessary to send samples of gravel away to be fire-assayed for gold. There are two good reasons for this. First, it is a practical impossibility to secure a small sample that will be even remotely representative of the whole deposit, or any substantial part of it. Second, the fire assay may reveal gold that cannot be saved by ordinary methods, and so give an overvaluation to the gravel. Such an assay will show *all* the gold in the sample, including that locked up in the sulphides and quartz particles, as well as the flour gold that will probably not be saved by ordinary riffles and concentration methods. If an experienced man, working carefully with the pan, cannot save more than a certain percentage of the values, it is a certainty that no better results will be obtained in the sluice boxes, or other larger-scale machinery, so expectations of a higher recovery, based on assay alone, are apt to bring disappointment in subsequent operations.

Character of gold in placers

Size, character, and fineness of the gold in the gravel will have a bearing on subsequent operations and recoveries. Coarse gold, that is retained on a 10-mesh sieve (diameter of particles greater than 1/16 inch), offers no difficulties in recovery. Any concentrating device or any system of riffles will serve. It is with the very fine sizes that trouble starts. Clean, bright gold is easily amalgamable, even in the fine sizes; but if the surface is stained with a film of iron oxide, or some other contaminant, the fine sizes have a tendency to float, and are amalgamated only with difficulty. Such gold is called "rusty" or "coated"; it often occurs in small flakes, very thin and flat, coppery in color, with little or no lustre, or it may even be almost black. This subject is discussed further in Chapter III.

Fineness of gold

The fineness of gold is the proportion of the pure metal in parts per thousand of alloy. Pure gold, 1000 fine, is worth $35.00 per troy ounce. Placer gold varies extremely in fineness. It is always found alloyed with some proportion of silver, sometimes with copper; and it may vary from 600 to more than 900 fine, in value from $21 to $35 an ounce. A variation of this extent, of course, has an important bearing on profit. Most placer gold is about 800-900 fine, and worth from $28.00 to $31.00 per ounce. Placer gold of low fineness can generally (but not always) be recognized by the lighter color, due to the silver alloyed with it, as well as by its streak on a streak-plate. After one becomes familiar with the gold in any district, he soon learns to judge the fineness with considerable accuracy according to its color and lustre; but what looks like gold of high purity from one district may prove to be quite the opposite in another. Very fine gold is often of high purity, which compensates to some extent for the trouble in saving it. This no doubt follows from the fact that the alloying metals are dissolved out more easily in the fine sizes after long immersion in water. Fineness of gold is best determined by fire assay, and it is always reported in mint returns.

Water supply and dumping ground

Two other factors, water supply and amount of dumping room for tailings, are always to be considered in valuing placer ground. These are discussed in detail in the chapters on sluicing. Ample water will compensate in many cases for low-grade gravel, and if tailings can be disposed of by gravity, without danger of being carried down on another property, a higher valuation can be placed on the ground. Water rights must be investigated, to make certain that prior rights for irrigation or mining purposes do not exist, which would be violated if the water were diverted for placer working.

Other things to bear in mind are accessibility and nearness to roads and supplies, as well as possibility of obtaining experienced men for work on the job. Before any extended plans are made for operating, all questions of title and ownership of the ground must be gone over; if it is a claim, it should be properly located and recorded (Chapter XII); if the ground is being leased, the royalty payments should be definitely settled on, as well as such points as tailings disposal, etc. This is especially important in states having anti-débris laws.

Business sense in mining

The purpose of making a thorough examination of placer ground, and obtaining all the data possible on these and other points that suggest themselves, is to determine the amount of money that can be profitably invested in equipping it, which will return the original amount and a profit. Merely the fact that a certain machine or a certain method is known to work economically is not enough. There must be sufficient yardage in sight to justify the capital cost with amortization. The principle holds good whether it is a question of building a $25 rocker, or putting in a million-dollar dredge. If only a small yardage of gravel is in sight, it is folly to think of putting in equipment and improvements which may be efficient and economical, but which never can be paid for before the available yardage is exhausted. It is much sounder business to use such equipment as will do the job, even at a considerably higher cost, if it will not involve a high initial investment for installation. For instance, a piece of ground is examined and measured up, and it is found that there are 1,500 feet of channel, 10 feet wide, with 2 feet of gravel above the bedrock that can be worked. This will be 30,000 cubic feet or 1,100 cubic yards. If the gravel runs 75 cents per yard, there is $800 to be recovered. This $800 is a limiting figure from which all expense of operation, including equipment, must be deducted before any profit is shown. It might be possible to put in a short sluice and shovel by hand, using wheelbarrows for transporting, or a much larger tonnage could be handled daily at a much lower cost by putting in a gasoline hoist and using a scraper or drag-line to bring in the gravel. Despite the lower per yard cost of the second plan, it would be poor business to do it, as the initial cost of the equipment would be disproportionate to the size of the job.

Dredging ground

A few words of general information may be added about dredging ground, though any detailed consideration is entirely beyond the scope of this book. To be suitable for dredging, gravels must be extensive in area in a wide valley, in depth 20 to 60 feet, with a soft and easily decomposed bedrock that can be cut by dipper or buckets. The bedrock must lie flat, or nearly so. Gravels must be free from large boulders that cannot be handled by the dredge. Most dredging properties have so low a gold content in the ground that they cannot be worked by other placering methods, and gravel with as little as 10 cents per yard, under favorable conditions, can be profitably dredged. Determining the value of dredging ground calls for the sound judgment of an experienced engineer.

Hints for prospecting

The inexperienced prospector, before starting out, will do well to gather from every source all the information he can about the country he intends to investigate. If he has in mind the known placer districts of the western states, he can obtain for a nominal amount detailed maps and reports from both government and state bureaus that give useful information on geology, past production, position of auriferous streams, roads, nearest supply points, etc. If time and opportunity permit, the files of the leading mining journals can be consulted at the large city libraries for additional information. Many states, through their mining bureaus, are especially helpful and give practical information that may save much time. The bibliography in this book gives a list of the best-known references on placer mining, as well as recent publications by the various states on the subject.

If placers are being prospected outside of the United States, the Bureau of Mines at Washington can frequently be of service, especially with information on the rights of American citizens engaged in mining in foreign countries, as well as on the mining laws of the country in question.

Equipment for prospecting

Equipment for preliminary prospecting is simple, and should be kept as light as possible for much travelling around on foot in rough country. Outside of camping supplies, it will include a light miner's pick for dislodging boulders, digging into potholes, trenching, etc.; a No. 2 round-point shovel with a spring handle; and the regulation miner's pan. A couple of strips of half-inch iron, about 18 inches long, shaped into a hook and spoon, are useful for gouging out crevices and seams in the bedrock, as well as a stiff brush for scraping fine dust into the pan. A good horseshoe magnet for removing black sand from the concentrates should be included, and a strong magnifying glass is needed for examining black sand for fine colors. A small bottle of mercury, a few ounces are enough, may be required for amalgamating the fine gold, along with a small pocket scale, with weights for weighing the gold, and a wide-mouth bottle with screw top to contain the gold and concentrates from each panning.

Best places to prospect

First attention is generally given to the stream beds and bars of creeks and gulches along their entire course. Such streams as have a moderate gradient, about 30 to 50 feet to the mile, offer better chances than those with a swifter current. There is little likelihood of finding concentration in the narrow gorges where the heavy rush of water will have carried along the gold. Wherever the valley opens out, benches of gravel on the hillsides and slopes should be looked for and tested. In most cases, the gravel on the quieter sides of banks and bars and the inside of bends offers the most promise. In a case reported from Montana, the gravel at the water's edge showed 6 or 8 colors to the pan; 15 feet back there were 25 to 30 colors, and 30 feet from the shore line 100 finer colors.

Bedrock, in every case, is the datum plane for the prospector. A good prospect can easily be missed by testing only the lean top gravels and quitting further work at that point to go elsewhere before reaching bedrock. Be sure that it is true bedrock that is reached; not a false bedrock or hardpan that may overlay the richer gravels below. Holes should be sunk at right angles to the course of the stream, some distance apart, and by taking pan samples from each one of about the same weight of gravel to test, and using the same care on each, a fair quantitative idea can be formed of the relative value of the ground, and some notion gained of the position of the pay-streak. Once the pay-streak is found, further work of course should be concentrated around it.

Bedrock crevices

Wherever bedrock is exposed, it should be carefully examined for depressions and potholes that may hold a rich pocket. If a crevice crosses the bedrock at an angle to the stream flow, it is usually worth prospecting with pick and crevicing irons as deep as it is possible to dig. Such a crevice, indeed any change in the bedrock that interrupts its normally smooth surface, acts as a natural riffle and permits concentration. It is amazing how fine gold will work its way down into extremely narrow cracks. Fractured bedrock should be entirely removed until solid material is reached.

In the formation of placers the coarse heavy gold, the nuggets of the 'forty-niners, were deposited along with the heavy gravel. Both dropped to the stream bed bottom when there was any slackening in the velocity of flow, while the smaller gold particles were carried along with the lighter gravels and sands. Consequently, if the pannings show a large amount of sand and very little gravel, the chances are that any gold present will be scanty in amount, and so finely divided as to be difficult to save.

An excessive amount of black sand in the gravel shows that conditions were favorable for deposition of gold as well, but it does not follow, unfortunately, that gold will necessarily be found associated with it.

Re-sorted placers

Because a piece of ground has already been worked over and yielded considerable gold production in the past, there is no reason to condemn it without a few tests, though undoubtedly the cream has been taken off. In the intervening years the gravels have been re-sorted by stream action, and new bars and benches have been formed, with re-concentration of the gold overlooked by earlier miners. These are worth testing for new values, as well as for what might have been over-looked by the old-timers. Careful prospecting is slow, patient work, involving considerable laborious digging, trenching, and sinking; but a systematic procedure, with application of a little knowledge of the local geology, is certain to result in the discovery of any gold that may be present.

CHAPTER III.

PANNING. PROPERTIES OF GOLD AND BLACK SANDS. ESTIMATING THE VALUE OF COLORS.

Panning as a means of saving placer gold is as ancient as the Pyramids. It is the poor man's method of working rich ground, requiring no treatment but his labor and the few necessary tools of the trade. Its economic importance in adding to the production of placer gold of the country is not large, because as a gold-saving device it is only adapted for rich gravels, and those in limited amounts. However, panning is used in every branch of placer mining for cleaning-up concentrates, as well as for preliminary testing of ground to see if it is worth working; and skillful use of the pan marks the good prospector everywhere. Simple as the operation is, it requires considerable practice to become an adept, and it is much easier to get the idea from watching an experienced miner manipulate the pan than from a description of the process.

The gold pan

The standard gold pan is pressed from a single sheet of stiff iron or steel, and turned over a wire on the edges to strengthen it. The larger size weighs 1½ or 2 pounds and is 18 inches in diameter at the top, with sides sloping down about 30 degrees to a bottom diameter of 10 inches. The depth is 3 inches, and it will hold about 18 to 20 pounds of gravel, or more than a heaping shovelful from a No. 2 round-point. It costs about 75 cents at supply houses.

A smaller 10-inch pan is frequently used in testing, which will hold 3 to 5 pounds of gravel. An ordinary frying pan, with the handle cut off and the bottom scoured free from grease, will also serve for such work.

For cleaning-up black sand concentrates with much fine gold, and for delicate work on flour gold, a copper-bottomed pan with steel sides can be employed. The copper is amalgamated with mercury, as described later, and the fine gold is caught on the amalgamated surface. This pan should never be employed for ordinary work, as the amalgam would be scoured off.

The batea

In tropical countries the natives use the batea instead of the pan, and are extremely skillful with it. Shaped something like an old-fashioned wooden chopping bowl, it is cut from a solid block of wood from about 15 to 30 inches in diameter, with sides ⅝ inch thick. In use, the batea is floated on the water surface and the gold and black sands concentrated by a peculiar rocking, swishing motion, while the water enters over one edge. A 22-inch batea will hold 1/3 cubic foot of gravel, and a skillful native can handle 80-120 cubic feet per day. It is claimed to be far superior to a pan in the hands of an experienced man.*

The trough washer is occasionally found in the West and South. It is a hemicylindrical trough in form, and is rocked on its curved bottom with a handle. The two ends are closed, with plug holes on one side near the top. In the bottom are two longitudinal riffles. Gravel is shovelled in, water added, and the

* J. L. Brady, Philippine Islands, personal correspondence.

device rocked until the material is disintegrated, when the plug holes are opened and the trough tipped until the muddy water runs out. Large pieces of rock are raked over the side, and this is repeated until the water comes out clear. Then, through a jerky motion, the light sands are washed over one riffle and the heavy sands and gold are left between the riffles. The latter are separated by panning.

How to use the pan

The pan is filled nearly full of gravel and placed under quiet water deep enough to cover the pan and its contents. While one hand steadies the pan, the other stirs the gravel, breaking up the lumps and disintegrating them until every piece is wetted. If any clay is present, it must be puddled until it is entirely dissolved. As the water gets muddy, it should be replaced with a fresh supply—hence the advantage of being near a running stream.

When the material is thoroughly disintegrated in the water, the large stones and pebbles are thrown out, the pan is grasped with both hands on opposite sides and, while still under water, is given a vigorous oscillating, circular motion, rapidly alternating, so that the contents are shaken from side to side. Through a classifying action, the heavy particles are given a chance to settle. The pan is then raised out of the water and tilted slightly forward, not to exceed 30 or 40 degrees, so that the lighter sands now in suspension are washed out over the lip of the pan. At the same time the lip is dipped into the water and quickly withdrawn; this washes over more of the light material. It is a good plan to give the pan a few vigorous blows with the hand or a block of wood, as this will help to settle the gold particles. Repeat the operations until nothing is left but a few ounces of concentrated material at the bottom, consisting chiefly of heavy sands and the gold, if any. Wash this down to a small amount of black sand and the gold, but do not try at this time to make a complete separation of the two. Instead, with a little clean water wash both sands and gold into another pan or a receptacle to dry, and start panning more gravel in the meantime.

By using clean water at all times, better work will be done; where clean water is lacking, it will pay to go to some pains to settle the water in a sump, as muddy water tends to float off the fine gold, particularly if it is in leaf form. For the same reason, the pan must always be *absolutely* free from oil or grease.

The amateur can get some practice in panning by taking some lead shot, iron filings, or other heavy fine material and after mixing them with ordinary sand and gravel, endeavoring to recover them by panning. When he succeeds in keeping them all in the pan, without losing any in the tailings, he can feel that he will have little trouble in saving coarse gold at least, which has a much greater specific gravity.

Using a magnet and blowbox

When a fair quantity of black sand and gold particles have been accumulated, and the material has become thoroughly dry, the larger colors can be picked up with some sharp-pointed tweezers, and most of the black sands—all that are magnetic—can be removed with a horseshoe magnet. In using the magnet, it is a good plan to wrap some cellophane around the ends; then, when the magnet is withdrawn, the sands will drop off without sticking to it. Such non-magnetic sands as still remain can often be removed by using a "blowbox." This is made from a sheet of heavy tin, about 8 inches long, 6 inches wide on one end, and 5 inches on the other. The edges are turned up about 1 inch on three sides, leaving one side open, and the corners soldered together. Put the sands and gold dust in the box, hold it level, and blow lightly across it, tapping the box at the same time. With some practice, a surprisingly clean separation can be made, the lighter sands being blown out and the gold dust remaining behind.

Using a copper-amalgamated pan

The use of a pan with amalgamated copper bottom for saving fine gold has already been referred to. To prepare it for such work, the bottom must be carefully cleaned, scoured, and dressed with dilute cyanide solution until the copper surface is bright. Mercury is then sprinkled on the surface and rubbed in until the hard, dry appearance of the copper is changed to a bright, moist condition, when it is ready to be used. The gold, if in amalgamable form, will be readily caught, and the sands can be discarded. The amalgam can be removed by a hard rubber scraper.

Instead of rubbing with mercury on the copper bottom as described, a perfect coating can be obtained by dissolving some water-soluble mercury salt, such as mercuric chloride, in a small amount of water and pouring the solution in the pan. The mercury will then deposit on the copper. Or mercury can be dissolved in dilute nitric acid and the solution poured on the copper, when the latter should be coated with it. After the solution is poured off, additional mercury can be added, which will then take hold of the copper. These methods are suggested by W. W. Bradley, State Mineralogist of California.

Amalgamating gold with mercury

Another method of recovering the fine values is to put about a teaspoonful of mercury in the pan, and stir the gold and sands vigorously with a piece of black iron so that the mercury has a chance to make contact with all the gold particles and form an amalgam, which can be saved without difficulty in subsequent panning. It is not so easy to save the surplus mercury, which can be decanted off and collected in another dish, although there is always some loss. Use of mercury in panning is not always advisable; it is very difficult to handle, particularly for the inexperienced, and bought in small lots it is expensive. If the fine and flour gold can be saved in no other way, however, and are in sufficient amount to justify the extra cost of saving them, then it is certainly worth trying.

How much gravel can be panned per day?

The placer miner who is obliged to rely solely on the pan for treating the gravel must realize that it will be *only* the exceptionally rich gravel that will yield him day's wages, and he must therefore try to find pay-dirt that is far above the average. Just a few colors to the pan, however consistent, will not pay out, because it is not physically possible for him to put through unaided enough yardage by panning alone, though the gravel might pay handsomely by some more ambitious plan of working.

A good miner, doing a careful job of panning, can handle about 6 pans per hour, or 60 in 10 hours. An expert, or one thoroughly familiar with the ground, working in sandy gravel without much clay or hardpan, might do double this. If much clay is present, speed will be very much cut down. Sixty pans are equivalent to about 1/3 of a cubic yard; under the most favorable conditions a yard a day is just about the top limit for panning. As gravel weighs from 3,000 to 3,400 pounds per yard, this is equivalent to handling over $1\frac{1}{2}$ tons of material. At Fairbanks the miners figure roughly 7 pans to a cubic foot of gravel, or 189 to a yard.

To make $2.50 or $3 by panning, handling about a half yard a day (70 to 80 pans), the gravel must evidently yield about $3\frac{1}{2}$ cents per pan. This would be equivalent to $6 per yard, which would be considered rich ground anywhere. It is unlikely that there is much of such ground open to location in this country that has not been discovered.

Properties of gold

The distinguishing characteristics of gold as compared to minerals likely to be associated with it in placers generally permit easy identification after a little experience. These unique qualities are its characteristic yellow color, its great weight, its ductility, and its chemical resistance to acids and alkalis.

Although yellow gold is proverbial, its color in placers varies from a light yellow, due to a native alloy with silver, to the dull, coppery, lustreless hue seen in rusty gold. Some gold is nearly black in color, owing to manganese stain, but this film is quickly rubbed off by slight abrasion and rarely confuses.

The weight, or specific gravity, of gold is 19.3 times that of water, as compared with a specific gravity of only 2.6 for quartz or most gangue materials. However, placer gold may vary from 15.6 to 19 on account of impurities. Even with the lower figure, it is more than six times as heavy as quartz, which explains its easy separation in any concentration process.

Its remarkable ductility allows it to be hammered out into the thinnest sheets without breaking, and explains the frequency of float gold in thin scales and flakes in placers and river bars. In the form of these flakes, gold floats rather easily, despite its great weight, especially if there is clay in the water, or a little oil. It has been estimated that a single cent's worth of gold (about 1/4 grain) could be hammered out and divided into more than 2,000 separate pieces, each recognizable as a color.

Gold is resistant, for all practical purposes, to ordinary acids and alkalis. It is dissolved with difficulty in aqua regia, a mixture of hydrochloric and nitric acids, but is not affected by either alone. Dilute solutions of sodium and potassium cyanide will dissolve gold if sufficiently fine, a fact taken advantage of in the cyanide process.

Minerals associated with placer gold

About the only common minerals with which gold can be confused in the pan are pyrite, the yellow iron sulphide, and yellow biotite mica. Pyrite (specific gravity 5) is lighter than gold, and is found along with the black sands, while gold lags behind in the ribbon of concentrates. It is brittle, and can be crushed into a brownish powder. Mica is very light and easily broken, splitting into scales. The color of gold is so characteristically its own that after a little experience no other test is needed to identify it.

Black sands

Numerous other minerals are associated with placer gold and recovered in panning. The commonest of these, generally called "black sands," are particles and grains of magnetite, the black magnetic iron oxide, and ilmenite, a well-known iron titanium oxide, somewhat similar in appearance. Along with them are a dozen other less common minerals. All are characterized by hardness and specific gravity about twice that of quartz, but very much less than gold. Magnetite grains can be removed without difficulty by a horseshoe magnet; ilmenite is only slightly magnetic, however, and must be washed away from the gold, unless mercury is used for amalgamating the latter. In amount, black sands will vary from 10 to more than 70 pounds per yard of gravel. Fine gold is nearly always associated with them when the black sands come from a gold-bearing placer, and it is difficult to separate the two entirely. Most of the values are found with the sands less than ⅛ inch (3 milligrams) in size. Means of separating gold from black sand are described in Chapter IX.

If black sands are in any amount, it may be worth while to have them assayed for gold content, and if sufficiently high, they can be sold to smelters. Of course, there is little use in collecting a few hundred pounds of them, unless other operators can be persuaded to make up a joint shipment.

Estimating size and value of colors

A "color" has no exact meaning. It is used indiscriminately by miners to refer to a small piece of gold less in size than a nugget, which again can vary from 1/16 inch upward.

It is extremely important to be able to estimate the size

and value of colors in the pan, so as to put some value on the ground that is being tested. Not only the number of colors to the pan, but also their size, thickness, and purity must be considered. Obviously, ability to make a good estimate comes only with experience, and even old-timers make bad guesses. The only really safe way is to pick out the colors from a counted number of pans, weigh them up on a pocket scale, and calculate the value of the ground. For instance, if 2 grains are obtained from 5 pans, allowing 170 pans to the yard, the gravel would average about $4 per yard, taking gold at $35 per ounce, and allowing 6 cents per grain, a safe figure for average placer gold.

The pocket scale is sold by supply houses, with weights, for $4.50. Its use will prevent many a disappointment. Troy weights are universally used in this country for gold measurements. The system is cumbersome and unhandy; it should be remembered that a troy ounce is about 10 per cent more than an avoirdupois ounce. In the Latin countries the metric system, with the convenient gram and milligram, is used. Conversion tables are given in the Appendix of this book.

A grain of pure gold is worth about 7.3 cents with gold at $35; hence $\frac{1}{4}$ grain is worth a little less than 2 cents. In size this would be about the same as an ordinary pinhead. But if this pinhead were of paper thinness, instead of being spherical, as could easily happen with flake gold, it would weigh very much less, although it might appear on the pan to have the same size as a thicker piece. At the rate of 60 pans per day, there would be needed 3 good pinheads per pan to make $3.50.

Classification of gold particles by screen sizes

A scale of sizes, quoted from C. F. Hoffman is as follows:

Coarse gold (nuggets) which remain on a 10-mesh screen (openings 1/16 inch).

Medium gold (small nuggets) which passes 10-mesh and remains on a 20-mesh screen (openings 1/32 inch).

Fine gold, which passes 20-mesh and remains on a 40-mesh screen (openings 1/64 inch).

Very fine gold, which passes a 40-mesh screen.

The following tabulation will prove useful:

	Average number of colors per ounce	Approximate value per color, $35 gold
Medium gold	2,200	1.5 cents
Fine gold	12,000	1/3 cent
Very fine gold	40,000	1/10 cent

Flour gold is much more finely divided than this. An instance is presented in Chapter IX of gold from the Snake River in Idaho where over 1,500 colors were needed to make a cent in value. However, as a compensation, flour gold is usually of relatively high purity.

CHAPTER IV.

THE ROCKER:

ITS CONSTRUCTION AND OPERATION.

The rocker, or cradle, is by far the most important gold-saving device for the placer miner, and he ought to be thoroughly familiar with this simple and time-tested machine. It is easy to build, as it requires few tools besides a hammer and saw, and no skilled labor. It can be operated by one man, at least in the smaller sizes, though two men can do a more efficient job. Finally, it will handle from 3 to 5 cubic yards of gravel in 10 hours, or more than six times the amount possible by panning, and consequently it makes much gravel profitable to the miner that would be unpayable by panning.

The rocker varies widely in form and size, but in its essentials it consists of a box or trough mounted on two rockers which are set crosswise underneath the box, a screen box or hopper above the box for receiving the gravel, and an apron beneath the latter to catch coarse gold, and a riffle frame, or system of riffles, along the bottom of the trough to trap the black sands and fine gold.

Rocker dimensions and construction details

In size, the rocker varies in length from 15 inches to 9 feet, in width from 8 to 30 inches, in height from 10 to 70 inches. The smaller sizes are unsuitable except for exceptionally rich

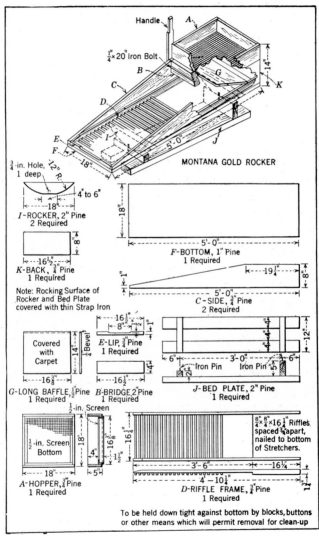

Fig. 3.—Rocker, with construction details. Adapted from Memoir 5, Montana School of Mines.

gravels, or for cleaning up concentrates. The usual size is $4\frac{1}{2}$ to 5 feet in length, 24 inches in height, and 18 inches in width. A rocker that is too short will prove very inefficient in saving fine gold. One that is too high means extra labor in shovelling into the hopper box and in lifting water.

Construction details are shown in the cut. Good lumber, if obtainable, is always desirable, free from knots and suncracks. Use 1-inch pine, and except for the bottom piece, dress it down to $\frac{3}{4}$ inch. The bottom should be a single board, not two pieces spliced or nailed together, and should be entirely smooth, as otherwise cleaning up between the riffles will be more difficult. The rockers, one cut from a piece of 2 in. by 4 in., the other from a 2 in. by 7 in., are screwed firmly to the bottom plank, care being taken that the screws do not project through the board. They should be supported on either side by cleats. This will give a 3-inch grade to the rocker, which on the average will be about right, but must be adjusted for the gravel and amount of water available for actual work. It is a good plan to flatten the rockers; this will give more of a bump when rocking, and help to disintegrate the gravel. In the cut, the rockers are shown of the same size and the grade is obtained by using crosspieces of different height on the bed frame. Either scheme is satisfactory.

The bed plate for the rocker is made of two pieces of 2-inch pine, 4 inches wide, about 3 feet in length, held by crosspieces. Each crosspiece has a $\frac{1}{2}$-inch hole bored in it, to take a spike that projects from each rocker. This latter is to prevent the rocker from sliding off on account of the difference in height. Instead of spikes, cleats can be nailed to the crosspieces for guides. The bed plate should be laid level, and anchored in place so as to be immovable.

Hopper box

The hopper box is removable from the rocker proper; it sits on cleats nailed to the sides, and should fit not too tightly, as the jar aids in settling. If it is 18 inches in cross-section, and 4 inches high above the screen, it will hold about 60-75 pounds of gravel. To the bottom of the hopper box is nailed a piece of punched plate of No. 18-gauge iron, perforated with $\frac{1}{2}$-inch holes, spaced about 1 to 2 inches apart. For fine gravel, No. 12-inch-gauge iron can be used. If the holes are counter-sunk, there will be less danger of clogging. Wire screen cloth netting with half-inch holes is sometimes substituted for the punched plate; it permits much more gravel to be treated, but is unsuitable if there is much clay in the feed, as the gravel does not get a chance to be sufficiently disintegrated. The cleats that support the hopper box should be set at an angle that will compensate for the grade of the rocker; if they were set level, the hopper would be tilted down hill when the machine was on the bed plate.

Apron

The apron is placed below the hopper box on cleats, set at an angle of about 40 degrees. It is a canvas-covered framework of $\frac{3}{4}$ in. by $1\frac{1}{2}$ in. lumber, with the side pieces extending at the lower ends a little beyond the lower crosspieces, so as to provide a clearance for the gravel. The canvas is so attached to the frame as to leave a pocket or sag about an inch deep at the lower end, which acts as a receptacle to catch the coarse gold and heavy minerals from the hopper box. The apron should be easily removable, and from time to time its contents are washed into a pan for a final clean-up.

Instead of canvas, a baffle board is sometimes used, on which corduroy, carpet, or blanketing is tacked, and the gold is caught in the meshes, which are washed at intervals in a tub of water. When worn out, the material is burned, and the ashes panned for fine gold. The coarse gold, in such case, is recovered in the riffles below the bottom of the rocker.

Riffle board arrangement

The bottom board of the rocker has riffles to catch the fine gold and black sands. These are usually the type called Hungarian, made of inch-square lumber, spaced an inch apart, and extending across the full width of the bottom board, flush with sides. They should be tacked firmly against the bottom, but the nails should be driven so as to permit easy withdrawal for cleaning up. Another plan is to bore ¾-inch holes through one side of the rocker at the end of each riffle to help in the clean-up. A cork is inserted in each hole when the rocker is being operated. Still another method is to construct a riffle frame, with the riffles nailed to bottom of stretchers. The whole frame is placed on the bottom board, and held there tightly by wedges or blocks, until removed for cleaning up.

In another arrangement the bottom board was painted with a priming coat, and 18-ounce duck canvas was stretched tightly over it, with riffles ½ inch square spaced 9 inches apart. The canvas made the bottom watertight around the edges when the boards were drawn up by the tiebolts, and saved the fine gold as well.

Instead of having the entire bottom covered with riffles, it is often a good plan to have the lower half of the bottom covered with carpet or cocoa matting, above which is tacked a wire screen to hold it in place, and the upper half equipped with riffles. The individual operator must choose which arrangement gives him the best results, from the standpoint of least labor, greatest gravel capacity, and most satisfactory recovery. The same thing can be said for the grade of the rocker; it may need to be varied to fit the individual case. By watching the tailings, and noting whether any values are escaping, adjustment can be made to effect improvement, as will be noted later.

The rocker illustrated in the cut will require about 30 board feet of lumber. For a permanent job, the corners should be strengthened with angle irons, placed outside, and tiebolts run through the sides. It will weigh, when dry, about 80 pounds; and when wet, in operation, considerably more. Hence, although it can be transported from place to place by two men, it is a considerable chore for a single miner to move around and set, particularly in a rough country, such as is likely to be found in prospecting.

Knock-down rocker

For preliminary work, as well as for working small yardages of gravel, where one must move frequently from place to place, the knock-down rocker illustrated in Fig. 4 is useful. The construction needs little explanation. The material needed to construct it is:

End, one piece of 1 in. by 14 in., 16 in. long.
Sides, two pieces of 1 in. by 14 in., 48 in. long.
Bottom, one piece, 1 in. by 14 in., 44 in. long.
Middle spreader, one piece, 1 in. by 6 in., 16 in. long.
End spreader, one piece, 1 in. by 6 in., 15 in. long.

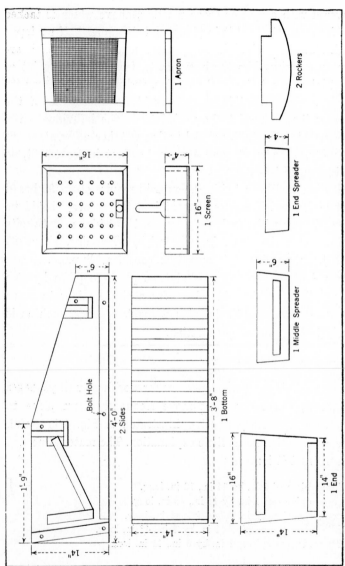

Fig. 4.—Knock-down rocker. H. H. Symons, California State Mining Bureau.

Rockers, two pieces, 1 in. by 5 in., 17 in. long.
Screen, about 16 in. square, with punched plate bottom. Four pieces of 1 in. by 4 in., 15¼ in. long for frame.
Apron, made of 1 in. by 2 in. strips, covered loosely with canvas.
For cleats and apron, etc., 27 ft. of 1 in. by 2 in., are required.
Six strips of ⅜-in. iron rod, 19 in. long, threaded 2 in. on each end and fitted with nuts and washers.

The rockers must be shaped as shown in the detailed drawing. Cleats are placed as indicated. When put together, the tiebolts are tightened and hold the bottom and sides firmly in place. The individual pieces must be cut accurately, so that the ends fit flush against the sides, as otherwise there will be leakage of water and escape of fine gold. If 1 in. by 14 in. boards cannot be obtained, clear flooring, tightly fitted, will have to serve, with additional cleats to strengthen. However, it will pay to get the single pieces. If the lumber is badly checked and cracked, cover with canvas, as suggested in the preceding paragraph.

The rockers are made the same height; hence, to get the proper grade, two planks about 2 in. by 8 in. by 24 in., placed firmly on the ground, one at a higher elevation than the other, serve as a bed for the rockers. Each must have a hole in the center to hold the rocker spike.

Rocker operation

If it is possible for an inexperienced man to spend some days with an old-timer, and watch him operate, adjust feed and water flow, get the correct slope, make a clean-up, etc., it will be worth pages of description. Treatment of every gravel varies in some details, and the best method must be worked out by "cut and try" operation. But in general this plan is followed.

The rocker should be set up near a water supply, and the hopper box filled about three-quarters full of gravel. Water is poured over the gravel with a dipper, unless some other arrangement has been made, and the machine is rocked vigorously, causing the smaller-sized material to pass through the screen and drop down into the apron. Water is added continuously in the meantime. At the end of each stroke make a short abrupt stop; this gives the gravel a chance to hit against the sides and the fine material is scoured off. If there is much clay in the gravel, it must be thoroughly puddled and the water increased, for most gold losses come from stiff, sticky clay that has not been disintegrated.

When no more fines come through the screen, and the gravel appears to be washed clean, it is examined for any large nuggets that could not pass through the holes; then the box is lifted out, the tails are disposed of, and another batch of gravel is rocked. After several hundred pounds have been put through, the apron is removed, and the coarse gold and the heavy concentrates, which should all have been caught in the apron, are dumped in a pan for a final cleaning later on.

Cleaning riffles

The riffles are cleaned up whenever it is thought necessary; not so often as the apron. Three or four times a shift should be sufficient. With the hopper clear of gravel, clean water is run into the machine for a few minutes while it is being rocked, so that all of the light sands and some of the concentrated material in the lower riffles are removed. The black sands and fine gold are brushed out from the riffles into a pan for further separation. If a blanket or corduroy has been used, it is carefully taken up and rinsed in a tub of water to recover the gold.

The riffles as well as the interstices of the cocoa matting should be watched to see that they do not pack with black sand, as this would permit the fine gold to slip over the top and escape. If black sands are present in quantity, increase the slope of the rocker. The proper grade is the least slope that will keep the black sands thoroughly agitated. The grade will vary with the rapidity of feeding, the amount of clay, and the fineness of the gold. Fine gold needs less slope than coarse gold, provided there is no packing of black sands. If there is trouble with the clay binding the gravel, the slope cannot be less than 2 inches; however, building the rocker longer will give a better chance for the fine gold to settle.

Water needed for rocking

The amount of water needed depends a good deal on the gravel. From 50 to 100 gallons per cubic yard of gravel treated are usually required, or 150 to 300 per 10 hours of average run. This is equal to 5 to 10 barrels. The water should come on the gravel as steadily as possible, not in a sudden gush, as this will carry away values over the riffles. Enough should be used to disintegrate the gravel completely, dissolve the clay, and carry the tails over the tailboard, and no more. Do not attempt to carry water to the rocker; it is more economical to do the reverse. If water is scarce, the discharged tails can be run into a sump or pit and allowed to settle; then the water can be drained off and re-used. Or a conserving box for water can be built of rough lumber, in size large enough to hold the rocker, and all rocking can be done within the box. As the clay settles it will stop up the leaks and the clear water can be drained off through a plug hole and re-used. As little as a barrel of water a day is said to serve with this scheme.

It may be possible with a little ingenuity to rig up a barrel above the rocker and have a steady flow of water to the screen box. This will give much more uniform results than applying with a dipper.

Ridding gravel of clay

If the gravel contains a considerable proportion of clay, it is a good plan to build a mud box or trough, into which the material is shovelled first and puddled around before going to the hopper. Otherwise beds of mud may form which will prevent settling of the gold on the rocker bottom. Occasionally, with much fine gold, mercury is placed in the lower riffles for amalgamating. In such a case, it is absolutely necessary that the water be clear and all clay dissolved, as amalgamation will not save gold smeared with clay. There is always a considerable loss of the "quick," but at times its use is fully called for.

Although a rocker can be worked with one man, much better results come with two men, particularly if water must be dipped up. A steadier operation results, and the rocker is not idle while concentrates are being cleaned-up. It is easier to move around and go elsewhere if the ground gets lean.

Steel rocker

The steel rocker shown in the cut is constructed along the conventional lines. It is 20 inches high, 19 inches wide, and

Steel rocker.

Rocking creek gravels in California.

4 feet long. Individual parts, which are detachable, are the rocker body, the rocking frame, perforated basket or screen, and the canvas apron. Its total weight is 126 pounds. It is manufactured by the New York Engineering Co., New York. List price is $70.

CHAPTER V.

GENERAL INFORMATION ON SLUICING.

Sluicing, in some form or other, has produced by far the greatest proportion of the world's placer gold. It entails some initial expense for construction of sluice boxes and introduction of a water supply, as well as preliminary work in preparing the ground for economic operation. It generally requires more than one man's labor, certainly so if more than one or two boxes are used. Its advantages are that it permits working far larger yardages of gravel per day than by hand methods, and hence a lower grade of gravel can be treated than would be profitable with pan or rocker. Also, the labor needed in sluice operation is considerably easier than in panning or rocker work, as water is made to do the work that was done otherwise by hand. Properly constructed sluice boxes, says Bowie, will save all the fine, floured, and rusty gold that can economically be caught, and the truth of this is shown by the failure to recover any values from the tailings of old sluicing operations when attempts to rework them have been made.

Since sluice construction means more or less expense and work in a single locality for some time, the first thing to be sure of is that the gravel contains sufficient values in adequate yardage to justify the entire expense of preparation, and return a profit over the expected cost of operation. This will have been determined by the methods suggested in Chapter II.

Almost as important as knowing that the gravel is profitable is to be fully assured of an adequate water supply that can be brought in economically, by natural flow if at all possible. Unless ample water is in sight, any plan to recover gold by sluicing has a hard, often impossible, road ahead.

Estimating water flow

Hence available water flow must be estimated in advance. Unlike water for panning and rocking, sluice water cannot be re-used, unless local conditions are extraordinary. Furthermore, instead of a few barrels of water daily, sluicing requires a minimum of about 5 gallons per minute for every cubic yard sluiced per 24 hours, or a total of 7,000 to 8,000 gallons. For instance, about 250 to 300 gallons per minute are needed to wash 50 yards in 24 hours. This is equivalent to the stream from a 3-inch centrifugal pump. If the gravel is clayey, or contains a considerable proportion of material larger than 4-inch size, a larger flow will be needed.

There are various methods of estimating water flow. If the water channel of the creek is fairly regular, the cross-section between the banks (width and average depth) can be obtained in square feet by measurement. The velocity of water flow in the stream is found by measuring the distance in feet that a chip will float downstream in a minute's time. Multiply these two figures and take 75 per cent of the product; the result will be approximately cubic feet of water per minute. To convert to gallons multiply by 7.5.

Water flow is variously designated in terms of miner's inches, cubic feet, and gallons. These are all convertible one into the other:

1 cubic foot of water = 7.5 gallons.
1 miner's inch = 1½ cubic feet per minute.
= 11¼ gallons per minute.
40 miners' inches = 1 cubic foot per second.

If it is planned to sluice but 10 hours per day, there must be enough water to treat 2½ times the contemplated yardage on the basis of 24-hour figures. For instance, if 50 yards are to be sluiced in 10 hours instead of 24 hours, about 600 to 700 gallons per minute will be required, instead of 250 to 300, which would be all that is required if the work were spread out over the longer period.

Grade for sluicing

In addition to an adequate water supply, the bedrock must have sufficient grade to carry along the water and the gravel at a rate sufficient not to choke the riffles, yet not so swiftly as to carry off the fine gold in the tails. The right grade is difficult to predict; it will depend on several factors—the character of gravel, whether sandy, cemented, or clayey; the size and shape of boulders, whether round or flat; the character of gold, whether coarse or fine; amount of black sand; kind of riffles employed in the boxes; and the amount of water available.

In most cases it has been found that a slope of 6 inches per 12-foot box (about 4 per cent grade) is about right for the average ground, but it may increase to 9 to 12 inches per box for pipe clay which needs a heavy grade to break it up. The same can be said for cemented gravels, which in addition need several drops along the line of sluice boxes for disintegration. Coarse gravels require 4 to 7 per cent grades and a proportionate increase in water. With transverse riffles such as

Hungarian that interrupt the water flow, a greater slope is needed than for longitudinal riffles (pole riffles). Similarly, with a given amount of water, more slope is needed for a wide sluice box than for one of the standard width (12 inches); and if grade is lacking, it may be necessary to use a box less than 12 inches wide to move the gravel along briskly.

It is essential that the grade of all sluice boxes following the first should not be less than that of the first, as otherwise there will be a slowing up of the water and gravel when the lower boxes are encountered, with clogging of the riffles and loss of fine gold. In general, there is less danger in having the slope too great than in having it too little, but if the grade is steeper than 13 inches to the 12-foot box, all but the coarse gold will be carried away.

If the grade of the bedrock itself is too slight to give sufficient drop for sluicing, recourse can be had to trenching, or to elevating the head boxes on trestlework. This latter, of course, involves additional labor for bringing in the gravel to the sluice. (See Chapter VII.)

Carrying power of water

The placer miner must know something of the carrying power of water. It varies enormously with the velocity. The following table* indicates the duty of running water at various velocities:

30 feet per minute	Fine sand lifted
45 feet per minute	Fine gravel moved
120 feet per minute	Moves 1-inch pebbles
200 feet per minute	Moves 2-3-inch pebbles
320 feet per minute	Moves 3-4-inch boulders
400 feet per minute	Moves 6-8-inch boulders

* Elements of Mining. George J. Young.

In every case it is essential that the water covers the largest piece of material to be moved.

Sluice head

Sluice head is a rather indefinite term used in placer mining to describe the volume of water needed to separate the gold from gravel in a sluice box. For a 12-foot box set on a 6-inch slope, it varies from 30 to 100 miner's inches (330 to 1,100 gallons per minute) to carry properly all the ground that 6 to 8 men can shovel in.

Average velocity of water in sluices of 6-inch drop to the 12-foot box is 140 to 400 feet per minute, which would correspond to a flow of 500 to 800 gallons per minute, depending on the depth of water in the sluice.

Frequently it is convenient to speak of per cent grade, instead of inches drop per foot, or per 12-foot box. The following table permits conversion:

Slope of 12 foot Sluice Box.

Inches	Inches per foot	Per cent
6	$\frac{1}{2}$	4.16
8	2/3	5.55
10	5/6	7.0

An instance of practical work is furnished from Young:[*]

			Water Flow		
Width of Sluice Box, Inches	Depth of Flow, Inches	Grade Per Cent	Cu. Ft. per Min.	Equiv. Miner's Inches	Cubic Yards Gravel per 24 Hours
10-12	6-7	4.1	45	30	67-135
12-14	10	6.2	100	66	150-300

This indicates that the duty per miner's inch (quantity washed in 24 hours) varies from 2 to 4 cubic yards of gravel, depending on the various factors listed.

[*] Elements of Mining. George J. Young.

Number of sluice boxes needed

The number of sluice boxes required (length of sluicing) depends on several factors. If the gold particles are coarse to medium size, 3 to 6 boxes (36 to 72 feet) will catch 90 per cent or more of the values, and in all probability, 75 per cent of the gold will be recovered in the first box. If the gold is fine, a longer string of boxes is needed. It is unwise to go to the expense of numerous boxes for a small operation, until panning of the tailings shows that gold is escaping which might be saved by additional lengths. In such a case, it may be necessary to change the form of the final box, and use an undercurrent (q.v.), or to adopt a different arrangement of the riffles. Very frequently a long sluiceway must be built, not for additional gold saving, but for conveying the tailings away to a dumping place. Gravel that is sandy and easily disintegrated requires only a few wide and shallow boxes to free the gold and permit its concentration behind the riffles; cemented gravel, and gravel with much clay, needs more boxes, with frequent drops between them to break up the material.

Sluice boxes should be laid in a straight line if possible. This is essential for boxes equipped with riffles; where the sluiceway acts simply to carry away the tailings, a slight curve can be used, and the grade increased. Sluice boxes of ordinary construction will last about 4 to 6 months. They are easily and cheaply constructed. Old sluice boxes should be burned, and the ashes panned for gold; it is worth while.

If it were possible to screen out oversize and worthless material before putting it in the sluice, the efficiency of the sluice would be very much increased. This is usually impossible, because all the gravel must be first washed to free it from fine coated materials. Similarly, if the feed to the sluice box is steady, not intermittent, much better work can be done. This, too, is frequently difficult, as the sluice is fed by shovel or wheelbarrow and the feed may come in sudden spurts that clog up the riffles and let the light gold particles ride over them.

Mud box construction

To avoid some of these troubles, it is usual to construct a mud box, as it is called, ahead of the first sluice box, into which all gravel is dumped. This mud box is an oversized sluice box, about a foot or so wider at one end and tapering down so as to fit the opening of the head box, with sides about 12 inches higher than the latter. The bottom is lined with pole riffles, or there may be a grizzly set across the box, made of heavy pipe or rail, half way above the bottom. The gravel is dumped on the latter and washed by the incoming water; oversize material and boulders are forked out and thrown to one side after having been cleaned. Clayey material is broken up and puddled in the mud box. Grizzly bars are set from 1 inch to 4 inches apart, depending on size and number of boulders and amount of water available. The mud box is usually set on a somewhat steeper grade than the succeeding string of sluice boxes; it may be 7 or 8 per cent.

Tailings disposal

Provision must be made for disposing of tailings at the end of the sluice, as they will stack up quickly. The best scheme, of course, is to run them into a rapid stream, unless local topography offers a dumping place. Sometimes a scraper can

The amount of gravel that can be sluiced per day varies so much with local conditions that an average figure can hardly be set. Sluicing is particularly adapted for gravel benches 5 to 15 feet in height. If ground sluicing is used and plenty of water is available, two men can handle 20 to 30 cubic yards per day, provided boulders are not too plentiful. Handling the gravel by hand (shovelling-in), a man can move about 5 to 10 cubic yards per day with gravel 4 feet in depth.

Cost of setting up sluices is small, as unskilled labor can make the boxes from ordinary lumber. Obtaining a water supply is another matter; it may be necessary to construct a dam higher up on the creek, and bring the water down by ditch, pipe, or even canvas hose. The points to consider in choosing a sluicing method are: (1) water supply; (2) grade of bedrock; (3) dumping room; (4) amount of boulders that must be handled; (5) amount of overburden. These, together with character of gravel and size of gold particles, will determine the most economical method of work, of the many different systems available.

CHAPTER VI.

SLUICE BOX CONSTRUCTION. THE LONG-TOM.
TYPES OF RIFFLES.

A sluice box is a three-sided open launder. In its simplest form it is built by nailing three boards together. It is usually made in 12-foot sections, widths from 6 inches to 3 feet, with 10 to 11 inches depth; but for customary small-scale work, the width is 12 inches and the depth somewhat less.

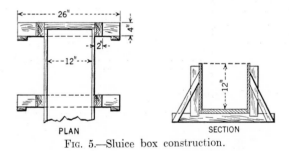

PLAN   SECTION

FIG. 5.—Sluice box construction.

Lumber chosen for sluice boxes should be planed on one side, and free from knots and cracks, in thickness 1 to $1\frac{1}{2}$ inches. The planed side is used for the inside of the sluice box. For permanent work use 2-inch lumber. Edges should be planed, and nails set fairly close to get a tight fit. The box must be watertight, for the gold will escape wherever there is a water leak. If leaks are observed, they must be caulked with wicking. Lining boards can be used on the sides of the box to take up wear. These can be ½ or ¾ boards and should be nailed from the inside of the box.

Three crosspieces, with uprights of 2 in. by 4 in. material, are required for each box to strengthen and prevent spreading of the sides. These are spaced every 4 feet and braced with 1 in. by 3 in. diagonals, as shown in the sketch. The crosspieces serve as foundations for the box, and where grade is lacking, rest on the trestlework for elevation. Instead of braces, the uprights may be nailed across at the top with a cleat, but this may interfere with clean-up, and should be avoided if possible. A neater but more expensive job can be done by mortising out the crosspieces to allow a recess in which to set the uprights.

Sluicing requires generally a "string" of 3 or more 12-foot boxes. The number needed depends on the character of the ground and the fineness of the gold, as described in Chapter V. Hence some means of joining the individual boxes is required.

Butt-end and telescopic sluice boxes

There are two types of sluice boxes—butt-end boxes and telescopic boxes. With the butt-end boxes, all boxes are the same size and interchangeable. One box butts up against the other, and at the junction an upright, spiked to each box, holds the two together. Ends should be sawed off square, so that there will be no cracks where the boxes come together. If desired, the sides and bottom piece of each box can be bevelled, in line with the water flow, so as to make a neat fit. The butt-end box is easily made, and the riffles can be set without difficulty.

The telescopic box is made by tapering one end of the box to 10 inches, and cutting the bottom board accordingly.

Thus the tail end of one box will fit snugly into the head end of another, and this is an advantage when the string of sluice boxes must be moved around from one place to another. Where the boxes join, gunny sacks, burlap, or old blankets are used to make the joint tight. Such material is placed under and up the side of the smaller end of the box, and the boxes pressed into place. The material should be burnt and the ashes panned for gold when sluicing operations are over, or when they must be discarded. It will contain considerable gold.

Telescopic boxes are good for small jobs, where it is known in advance that there is no great yardage to be treated and the boxes must be moved elsewhere. They have several disadvantages, however. They are harder to construct than the other type, and require careful dimensioning. It is more difficult to set riffles effectively than with the plain box. Their construction also interferes with the water flow, by narrowing up the cross-section at the end of each box, and so creating a surge that may carry off fine gold. Evidently there will not be a smooth line of flow as with the other type, as there will be a distinct drop in level where one box rests on the other. However, this may not prove a disadvantage, as it will help in breaking up the gravel when it is cemented or contains clay.

The cost of making sluice boxes is small, but bringing in 12-foot lumber over rough trails and inaccessible places is often difficult. About 50 board feet are needed per box. The cost per box, including labor, should not exceed $5 to $10.

The long-tom

A long-tom is a short, modified sluice, much used in placer work where there is not enough water for a string of sluice boxes, or where lumber is lacking, or where there is not sufficient grade to carry off the tailings. The capacity, of course, is much less than that of a string of sluice boxes, but it is well adapted for working a small area of gravel when there is sufficient water but too little for sluicing. Much more water is required than for operating a rocker. In 10 hours, two men can handle 3 to 6 cubic yards of sandy gravel with a long-tom, or somewhat less of cemented gravel. Properly regulated, it does good work on both coarse and fine gold and is espe-

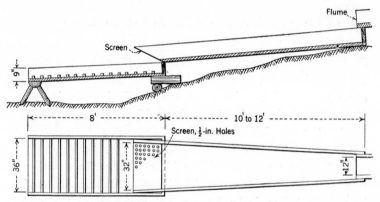

Fig. 6.—Long-tom, section and plan.

cially used for working gravel that contains much clay or is slightly cemented.

The long-tom is built in two sections, as shown in the cut. The sluice box, 6 to 12 feet in length, is 15 to 20 inches wide at the upper end, flaring out to 24 to 36 inches wide at tail or lower end, with sides about 8 inches high. At the lower end a screen or a piece of punched steel plate with ⅜- to ½-inch holes is set at a 45-degree angle as shown. This serves to prevent the coarse gravel from going on to the riffle box, which is the second section. This is usually shorter than the sluice box, usually about 8 feet in length, and is slightly wider than the other at the tail end. It receives the water and all the fine material that passes through the screen.

The slope of the sluice box is about 1 inch to the foot, or 12 inches to the 12-foot box, which is about twice the drop used in regular sluicing. The riffle box is set at somewhat less slope than the other. It is lined on the bottom with canvas or corduroy to catch the fine gold, which should be secured so that it is removable for washing in a tub. Cross riffles are placed across the bottom; these similarly must be removable for cleaning-up.

It is best to build the upper section of 2-inch lumber instead of 1-inch, as it must stand much abrasion of the gravel and much shovelling and puddling of the material.

Operating a long-tom

In operation the gravel is shovelled into the sluice box, into which a steady stream of water enters at the head from a flume or pipe. The gravel is worked over by a rake or fork to break up the clay and disintegrate the material. The fine material goes through the screen; the coarse is shovelled out and discarded. It is necessary, of course, to keep the screen from clogging. Clean-up is made, when required, by removing the riffles, collecting the gold and black sand, and panning. Sometimes mercury is introduced in the lower riffles to amalgamate the fine gold.

Riffles and their functions

Correct and efficient riffle arrangement is most important in gold-saving in rocker, long-tom, or sluice box. Riffles have several functions. Primarily, they form pockets in which the gold may be retained and later recovered. They retard the gravel and sand moving over them, and so give the gold a chance to settle. On account of their shape (pronouncedly with Hungarian riffles) they form eddies or "boils" which give a rough classification to the material in the riffle spaces.

Pole riffles

There are many types of riffles, and each has its defenders. Pole riffles are perhaps the commonest. These can be made from 3- to 4-inch peeled saplings, which can be cut right on the ground. They are sawed 3 ft. 8 in. long, with both ends squared, and all knots removed. This allows three sets to the 12-foot box. Two-inch strips of lumber, the width of the sluice box, and in depth equal to the diameter of the poles, are nailed to each pole, leaving about ⅜- or ½-inch distance between poles. The poles must be flush with the bottom of the sluice box, and each set of riffles butts against the other, as shown in the sketch. The crosspieces can be wedged into the sides of the box, or nailed from the outside, so that the nails can be drawn and the riffle-sets removed for cleaning-up. The crosspieces, which get heavy wear from the gravel, can be protected with strips of flat iron screwed along their length. Holes should be countersunk.

Pole riffles are good for saving coarse gold, but are not so efficient for fine values. They do not disintegrate the gravel so well as other types, as there is less drag on the flow. Sluice boxes with pole riffles will handle the maximum amount of material with the least amount of water. They wear out quickly and must be replaced, but this takes little time. With much clay, 2-inch squares of 1/16-inch iron plate can be hammered cornerwise into poles, in staggered rows, to act as knives and break it up.

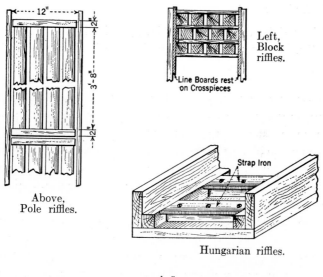

Above,
Pole riffles.

Left,
Block
riffles.

Line Boards rest
on Crosspieces

Hungarian riffles.

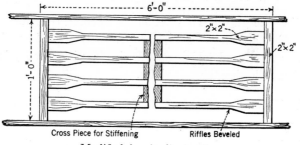

Modified longitudinal riffles.

Fig. 7.—Various types of riffles.

An interesting type of longitudinal riffles is sent from a small Arizona operation, described on p. 81. These riffles shown in Fig. 7, had less tendency to clog than the transverse type.

Block riffles

Block riffles are cut from 2 in. by 2 in. lumber, in 3-inch lengths. Lumber used should be long-grained, such as pitch pine, the kind that "brooms" on the end after wear, rather than hard wood, which wears round and smooth. The blocks are nailed to 1-inch square strips which extend across the width of the sluice box. About ¾-inch should be left between each block, and the blocks in the second and following rows should be staggered to break up the water and gravel flow that comes through the spaces of the first row. (See sketch.) All blocks should be cut the same height. The block units are held firmly in place by placing line boards on the sides of the sluice box which rest on the crosspieces. Thus, when the line boards are removed, the units can be lifted out without trouble for cleaning-up. Block riffles produce considerable drag and "boiling," but they are not so effective for disintegrating the gravel. They require a steeper grade than pole riffles. They are subject to rapid wear, but they can be replaced easily and are good gold-savers. Their place is generally in the head box following the pole riffles.

Rock riffles

Rock riffles can be used, when suitable stones are handy, for the head box. Tranverse poles are nailed across the sluice box at 3-foot intervals, and flat rocks, placed on end, about 3 inches in height, are used to fill up the space. They have the advantage of being practically indestructible and are especially good for cemented gravels, but they require considerable time to remove and replace for cleaning-up, and must be carefully washed to free them from fine gold. They require more water and slope than other types of riffles.

Cross riffles

Simple cross riffles are 1 in. by 2 in., or 1 in. by 3 in. strips, in length the width of sluice box, which are inserted at intervals along the box, and made fast by nailing through the sides. They wear rapidly, and provide little "boil" for getting rid of the sands. However, they can be used in the lower boxes to advantage, and are generally found on rockers and long-toms. They should not be spaced too far apart. A variation is zigzag riffles, which do not extend across the full width of the sluice, and are nailed to the bottom of the box. This makes them hard to remove for clean-up, but they are said to be good for fine gold. Still another variation of the cross riffle is to saw a 1½-inch strip diagonally so as to give it a 1-inch bottom; this provides the needful eddying action, when set up in the box, with the concave side pointed downstream.

Hungarian riffles

The Hungarian riffle is simply the cross riffle with an overhang on the down side, made by screwing to the crosspiece a piece of strap iron 1 inch or 1½ inches wider than the crosspiece. Holes should be countersunk. Riffle rows should be about 1 inch apart. This provides a strong "boil" to the water flow, and is effective in saving the finer values. This type of riffle clogs more easily and costs more to put in than other types, but its use is advised wherever black sand and fine gold are prevalent. Quicksilver can be introduced in one of the lower riffles, if conditions require it, for gold recovery.

Riffles are nailed to stretchers 4 to 6 feet in length, and these are held firmly against the bottom of the box by the line boards.

Floor linings

For the lower boxes it is a good plan to line the bottoms with cocoa matting, burlap, or corduroy for catching any fine gold that may have escaped the riffles. Such material should be protected with a fairly heavy wire screen, $\frac{3}{8}$- to $\frac{1}{2}$-inch mesh, and easily removable, so that it can be taken up for cleaning at proper intervals. At such times, rinse thoroughly in a tub, and pan the concentrates that are washed off for the gold. All old cocoa matting, corduroy, etc., should be burned and ashes panned, as with riffle blocks. If only a part of the sluice box is lined with matting, it should be set in a slight recess in the bottom, so that the sands can flow over it easily, without piling up. Special care should be taken to see that the interstices of the cocoa matting do not fill up completely with sand, as in that case the fine gold will ride over without a chance to be caught in the meshes. More water, or greater slope, will keep the matting clean. If there is too much material for the matting to handle, a $\frac{1}{8}$-inch steel plate with $\frac{1}{8}$-inch holes can be set on wooden longitudinal strips which rest on the matting. This will keep the oversize material away from the matting, but permit the sands to drop through. This arrangement, called a caribou riffle, is described with a cut in Chapter VIII.

Rubber riffles

Some success is claimed for rubber riffles. These are made from a special soft rubber, in sections 18 inches long by 12

Sluicing river gravel in California.

A short sluice for bench gravels.

Small sluicing operations.
Photographs courtesy of W. W. Bradley.

inches wide, of half-inch thickness. They are placed on the bottom of the sluice box and held down with a line board or strip. The manufacturer, V. L. Holt, Portland, Oregon, states that a heavy grade (8 per cent) and plenty of water are required. Several types are in service, one with 2,000 cells of 9/16-inch holes to the section, another with 1,000 cells of ⅜-inch holes. The water swirling in these cells keeps the sand and gold moving, with the gold particles sticking to the concave walls of the rubber cells, while the upward movement of the water carries away the sands. It is also claimed that these rubber riffles can be used in ground sluicing by laying them on the bedrock and protecting them with poles to keep off the large rocks. They can be easily cleaned by rinsing in a tub, the concentrates being recovered.

Number of riffles needed

The number of riffles needed depends on the richness of the gravel and the character of the gold. There should be enough to save all the gold economically possible, and no more. A practical check is always at hand by panning the tailings from the end box. If gold is escaping, put on another box, or change the riffles in the others to another type. Riffles must be carefully watched to see that they do not pack and clog; if they do, their efficiency as gold catchers has departed. Keep them loose. The eddies or "boils" must be strong enough at all times to prevent the riffles from filling up with heavy sand, though if the boiling action is too strong, there is danger of light flaky values being carried over. Therefore, the cocoa matting in the last boxes is a good safety measure.

Chapter VII.

METHODS OF WORKING PLACER GROUND.

The choice of a mining method for placer ground depends on such factors as amount of overburden, height of gravel, available dumping space, slope of bedrock, water flow available, size and number of boulders that must be moved, richness of gravel, size of gold particles, limits of pay-streak, amount of money that can be spent on equipment, and terms under which the ground can be worked. Each must be given consideration. Final choice will depend on what is best adapted for local conditions.

Ground sluicing

This method is adapted for shallow gravels, 6 to 8 feet thick, especially bench deposits, with an overburden of soil or unprofitable material that must be removed before reaching pay-ground. Adequate grade of bedrock is necessary; the slope should be not less than 4 per cent, more if possible. Enough dumping room must be available for tailings; with bench deposits, above stream line, this may not offer difficulties. Water must be ample; it needs about four or five times as much water for ground sluicing as for handling the same amount of ground by shovelling-in. It has the merit of requiring no expense for equipment, though considerable time and labor may be needed for constructing ditches and arranging for water, and much more gravel can be handled than by shovelling. Two men should handle 20 to 30 cubic yards per day by ground sluicing, at a cost of 15 to 35 cents per yard.

A trench is first dug to bedrock at the lower end of the deposit along the side, and a stream of water is sent through it, either from a dam above in the creek, or by flume or pipe. The overburden and as much of the gravel as possible are sluiced into the trench, by picking and shovelling at the bank, and directing the water flow so as to aid in the undercutting. Thus the lighter material is washed away in the tailings, while the concentrated gravel containing all the gold values remains behind for consequent cleaning-up. No sluice boxes

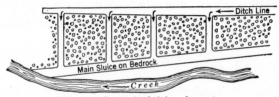

Fig. 8.—Ground sluicing layout.

are ordinarily required at this stage, as the depressions in the bedrock form natural riffles to catch the gold. If these are lacking, the bottom can be paved with cobbles, flat end up.

There are many variations of ground sluicing, which ingenuity will suggest. It may be possible to run a ditch from some upstream along the upper side of the gravel bench; the water is then diverted at right angles to the bench and falls down over the gravel bank, cutting channels, which can be caved. Smaller ditches take the material to the main sluice. (See sketch.)

Another plan is to use a plow or harrow to loosen the top of the overburden, then with a slip or drag scraper to convey the light material to the main sluice. Under suitable conditions with loose gravel a team of mules pulling a scraper will handle 30 to 40 cubic yards per day over a distance of 75 feet, at a cost of 30 cents per yard, about one-third the cost of handling the same material by shovelling.

Before abandoning any ground sluice, the bottom should be cleaned with unusual care, and the bedrock removed for several inches (particularly if creviced) for clean-up.

Booming

"Booming" is sometimes called "hushing." Where there is not enough water for ground sluicing a dam can sometimes be built a short distance up the creek for impounding the water and so obtaining a large flow for a brief interval. The dam is equipped with an automatic gate, which will release all the water suddenly when the reservoir is full. The ensuing rush of water will wash off the top soil and some of the gravel to the sluices. When the reservoir is empty the automatic gate swings back and it fills again. Such a gate, 4 to 10 feet wide, is built of heavy timber and swings outward over a horizontal pivot set one-third the height of the gate from the bottom. When the water rises to more than two-thirds the height of the gate it automatically opens.

An average-size reservoir should be about 10 feet wide, 6 feet deep, and about 100 feet long. It will contain about 22,000 gallons of water when full. Booming often proves a cheap and effective way of removing overburden. An instance is given of an area 900 feet long and 25 feet wide stripped of muck and gravel 5 feet thick in 3 weeks; the cost, including dam and gate, did not exceed 7 cents per cubic yard.

However, it is necessary to have a wide sluiceway for the rushing water and gravel, as otherwise the ditch will be immediately choked up, and fine gold will be carried off and lost. It is an ineffective way to recover gold and should, like ground sluicing, be considered a preliminary to shovelling-in.

Shovelling-in

As the name suggests, shovelling-in demands considerable manual labor, hence it is adapted for handling rich gravels or gravels previously concentrated by ground sluicing or booming. It generally follows these two methods of working placer ground. Sufficient grade is required for the sluice boxes; an ideal condition is where the bedrock slope exceeds that required for the sluice box. In such a case the lower boxes can be elevated on trestles, thus providing dump room. Where such grade is lacking, the head box must be elevated sufficiently to carry off the gravel. Water is needed in sufficient quantity to handle the yardage. (See Chapter V.) If there has been enough water for ground sluicing, there is usually ample for shovelling-in.

The work includes transporting the gravel to the sluice boxes, usually by hand shovelling, cleaning off the bedrock and frequently taking it up a few inches, moving boulders and cleaning them of fine material, disposing of tailings by scraping or shovelling unless they can be stacked, and cleaning up the riffles.

The line of sluice boxes may be set centrally along the length of the placer at the proper grade and the gravel excavated in longitudinal cuts, or it may be set close to the bank, and as the latter is cut away, the sluices are moved over. Another variation is to have secondary smaller sluices that run transversely to the main sluice and lead the material to the head or mud box.

With the first scheme, the string of sluice boxes can be set permanently, but shovelling cost becomes prohibitive, as the distance from the sluice boxes becomes too great to throw the gravel. The amount of gravel that one man can handle in shovelling-in depends on whether the gravel is loose or cemented, when it may require considerable picking, as well as on the distance to the sluice box that the gravel must be thrown, and the elevation of the box. Under average conditions, one man can handle 7 to 8 cubic yards in 10 hours, with a gravel bank 4 feet high and a lift of 5 feet. If bedrock is being cleaned, this capacity is much reduced. Splash boards are often set on the far side of the sluice box into which gravel is being thrown; these are movable.

Boulders, if of small size, are thrown on the grizzly of the mud box and washed clean of fine material; they are then raked off and thrown to one side. Large boulders are pried out and stacked on the cleaned bedrock, or mudcapped with a half stick of dynamite to break them. If boulders are too plentiful and too large, it may make the placer unprofitable.

Transporting gravel to the sluice boxes

When the distance from the sluice becomes too great for shovelling, the gravel can be transported to the sluice in wheelbarrows. Capacity of a wheelbarrow is about 11 to 13 pans, or $1\frac{1}{2}$ to 2 cubic feet of gravel in place. To strip 1 square foot of bedrock with 5 or 6 feet of overburden requires 3 barrows, at this rate. Self-dumping cars on light tracks can also be used for transporting. If the contents of the wheelbarrow or car are dumped into the sluice all at once, a mud box is needed, otherwise the rush of material will clog the riffles.

Cars can be run up an incline and dumped into a hopper with a small gasoline hoist. In some cases a drag scraper with double drum hoisting engine can be used effectively to bring the gravel to the sluices. The hoist is set near the mud box and drops the scraper over grizzlies set over the mudbox, into which it discharges. Snatch blocks and movable tail sheaves permit a large area to be excavated from a single set-up. Of course, any such installation as this runs into money, and should not be contemplated until it is definitely known from the work on a smaller scale that the values are there to justify the capital expense.

The cost of dams, drain ditch, water ditch, and 10 sluice boxes may run all the way from $500 to $2,000, depending on local conditions, wages, transportation, etc.

Hydraulicking

Hydraulicking is the cheapest method of working placer ground but is generally associated with large-scale operations, treating thousands of yards per day, and any detailed discussion is beyond the scope of this book. The reader is referred to several good treatises on the special subject for such information. (See Bibliography.)

However, even for the small operator, hydraulicking frequently has a place, if a good flow of water under head can be obtained cheaply. In none of the other methods described above is hydraulic head essential; but in hydraulicking, the whole efficiency depends on it. Sufficient grade of the bedrock is also required,

Design of pipes, flumes, and giants to handle high pressures efficiently and safely is an engineering job. A table showing what can be expected from small giants with 1-inch nozzles is given in the Appendix.

Cases are reported where shallow placers have been worked cheaply by hydraulicking, with heads as low as 35 to 100 feet of water. H. I. Ellis* gives an instance at Circle where a 9-foot gravel bed was worked with No. 1 giants (2-inch nozzle) under a 100-foot head, diameter of pipe inlet 7 inches, water flow 94 cubic feet per minute. Six 12-foot sluice boxes, 30 inches width, with block riffles, handled the gravel, at a cost of 12 cents per cubic yard. The bedrock was washed clean of gold without hand methods. About $2\frac{1}{2}$ cubic yards of gravel per 24 hours were handled per miner's inch. An instance is reported in Nevada of hydraulicking with a $3\frac{1}{2}$-inch hose with $\frac{5}{8}$-inch nozzle under a 50-foot head; and at Dahlonega, Georgia, the shallow bench gravels, sometimes only 6 feet high, are hydraulicked with water heads of 50 to 200 feet.

Hydraulicking with canvas hose

Canvas pipe hydraulicking has a place in small-scale work. This pipe, called sluicing pipe or flume hose, can be obtained from supply houses. It is light and flexible, but should be handled rather carefully to prevent chafing and abrasion over rough ground. It should be dried out after being used, and when installed, should be turned over frequently or set in boxes to prevent rolling. It comes in lengths of 18 to 20 feet, in widths of 24 inches. With a $1\frac{1}{4}$-inch lap, the diameter of the pipe is about 7 inches. Fresh pipe lengths are inserted by

* Engineering & Mining Journal, Vol. 99, pp. 805-810.

Loosening gravel and removing clay before washing in long-tom.

Small scale hydraulicking near Mountain City, Nevada.
Courtesy J. A. Fulton, University of Nevada.

telescoping ends into each other; this makes a tight joint under pressure. In small hydraulic operations the water can be led to the pit from the head ditch or penstock by canvas pipe of 6- to 10-inch diameter, provided the head does not exceed 50 feet or not more than 1000 gallons per minute are handled. Sometimes the outside of the hose is tarred to protect it. The nozzle is held in the hand when piping or supported on a tripod. Such an arrangement is makeshift, but it will serve for a time.

Successful hydraulicking invariably uses water under natural head to supply pressure. Wherever pumps have been used for this purpose the result, either in large- or small-scale work has been unprofitable, unless power was obtained at far less than commercial rates. (See table in Appendix, p. 139.)

"Blowing-down"

In Australia, shallow placers of low gold content have been worked by starting at the top instead of the bottom of the deposit, and hydraulicking down hill to the main sluice. This is called "blowing-down." The general arrangement is similar to ground sluicing, as illustrated in Fig. 7. After the main sluiceway has been driven to bedrock at the lowest point of the deposit, a side channel at right angles is taken up hill to the top, where hydraulicking starts. The gravel is carried down to the main sluice, where the gold is caught. It is claimed that this permits more effective cleaning up of the bedrock, as there is no tendency to drive the gold into the crevices. It is necessary to have enough grade for the side channels to permit easy removal of the gravel down hill, that is, at least 4 per cent.

Other methods of working placer ground

There is no set method of working placer ground. All that has been said before is merely suggestive of what has proved good practice under the right conditions. Ingenuity, common sense, "cut and try," are required for success.

It is not always necessary, in small operations, to have a long string of sluice boxes. Gold of medium size can be saved in almost any kind of box, with almost any riffle arrangement. Hence a single box will frequently save 80 per cent of the gold; what escapes may not justify the expense of trying to get it. The actual mining of the gravel, rather than the saving of the gold, is what demands the best thought and ingenuity of the operator.

Some instances of practical work on a small scale follow. At French Creek, Montana,* an 18-inch deposit on bedrock is worked through two telescoping 12-foot sluice boxes, 10 inches wide at one end, 8 inches at the other. Grade is 6 inches in 12 feet. Water is pumped from creek to height of 8 to 12 feet into the sluice boxes, with a 2½-inch gasoline pump, 2-inch discharge, with a gasoline consumption of 4 to 5 gallons per 10 hours. The sluice box has Hungarian riffles.

At another property,* a belt conveyor with cross cleats on the upper side elevates the gravel to the sluice box, the gravel being shovelled in from the pit bottom. A 6-horsepower gasoline engine furnishes water and power for the conveyor.

At a bench deposit,* sluice boxes were laid on the creek bed, and the stream furnished water from a dam. Gravel was taken by wheelbarrow from an open pit on creek bank and

* From Placer Mining number of The Black Hills Engineer, South Dakota School of Mines, Rapid City, South Dakota.

dumped into the mud box. Sluice boxes, three in number, were made of 2 in. by 12 in. material, with 12 in. by 12 in. cross-section. Riffles consisted of staggered rows of 1½-inch holes spaced 1½ inches apart bored in 2-inch plank. This type of riffle is popular in Montana, but is not reported elsewhere.

Instances of placer work in Nevada*

In Nevada, where water is generally scarce, use of trommel or screen to disintegrate the gravel and get rid of oversize material at once is quite common in placer operations. A typical instance is presented: Gravel is taken to a 12-foot revolving trommel, 4 feet in diameter, with ¼-inch mesh screen, where water, amounting to 50 gallons per minute, disintegrates the material. The fines are washed into two power-driven rocking sluice boxes, 16 ft. by 4 ft., with a grade of 3 inches to the foot, with transverse riffles. There is a short stationary sluice box below each rocking sluice with a grade of 1½ inches to the 16-foot box, 19 inches wide, with amalgamated copper plates. An amalgam trap is at the end of each box.

Another Nevada operator has a washing plant of shaker box and sluice. The shaker box is 5 feet long, 30 inches wide, 18 inches deep. Capacity is 16 yards in 8 hours. The oversize, larger than 8 mesh (about 3/32 inch) is discharged as waste; undersize goes to sluice, 10 inches wide, 8 inches deep, 48 feet long, with coarse wire screens for riffles. Cost of operation is reported as $1.50 per yard.

At another Nevada operation the gravel in a creek bottom is loosened with a horse-drawn harrow while water flows over it. This washes away the soil and clay. Gravel goes to a long-

* From Placer Mining in Nevada, published by the University of Nevada.

Gouging out the pay streak from a bench placer in Idaho.

A small shovel operation in California. (Photo W. W. Bradley.)

tom, power driven, set on rockers, 16 feet long, 10 inches deep, 12 inches wide, driven by an eccentric from a 1½-horsepower gasoline engine. Capacity is said to be 4 cubic yards per hour, with a good recovery despite the small water consumption. (See cut.)

A small Arizona operation*

In Arizona, a young engineer worked a gravel bank on the side of a small hill, 150 yards long by 30 yards wide, with a thickness of 15 feet, by building a ramp from the hill directly above the sluice box and sloping it so that the gravel dropped 5 feet to the grizzly, which helped disintegrate it. Gravel of egg size was carried by the sluice box, which had a drop of 6 inches in 12 feet. The box was 40 feet long with a 12 in. by 12 in. cross-section. In constructing the box, lampwicking was used between all joints to make it watertight, and it was lined as well with common black sheet iron. Between the overlaps of the iron he used "plastic wood" to prevent gold getting under the lining. This arrangement he recommends highly. Longitudinal riffles, he adds, were far more satisfactory for this job than either transverse or Hungarian type, and were arranged as in Fig. 7, p. 63. There was less clogging than with the others. Each box had 6 sets of riffles, each riffle being 6 feet long. Two sets, he says, would have been ample, as no gold was found below the second riffle, and 95 per cent of it was under the first 2 feet of the first riffle. The gravel averaged 30 cents per yard, and about 20 yards were worked daily, with a water consumption of 360 cubic feet per minute, with sluice box running about half full.

* Personal correspondence.

CHAPTER VIII.

CLEANING-UP. RECOVERY OF FINE GOLD FROM BLACK SANDS. UNDERCURRENTS AND SIMILAR DEVICES.

Cleaning-up the sluice boxes is done at more or less regular intervals, depending on the richness of the gravel, as well as upon other factors. Until one is familiar with the ground from actual experience, a clean-up should be made every few days. Afterward, clean-up may occur much less frequently. There is sometimes the danger of theft if left too long. As cleaning-up requires some time, during which all sluicing operations necessarily cease, it is good business not to do it more often than necessary, particularly in a short working season.

The longer the string of sluice boxes, the less danger there is of loss of gold that may escape the riffles in the first boxes and continue with the tailings to the dump, as the riffles in the lower boxes are always there for recovering it. With a short string, or a single box, there is no such safety, hence clean-up should be more frequent.

In by far the majority of cases, however, all the coarse gold will be found in the first box, and following boxes will show a progressively smaller amount in the riffles. An instance is given where 80 to 90 per cent of the coarse gold was recovered in the first three boxes, which were of the customary 12-foot length, set on 6-inch grade, with pole riffles. As another instance, Bowie states of Australian operation that with a 12-inch sluice, grade 1 in 48 (about 3 inches to the box), water flow 600 gallons per minute, 95 per cent of the gold was lying within 3 feet of where the gravel was shovelled into the sluice. The bottom boards, he adds, were perfectly smooth, yet a powerful current failed to move the gold.

Procedure in cleaning-up sluice boxes

When it is determined to clean-up, no more gold is permitted to enter the boxes, and clear water is run through for sufficient time to clean out any surplus feed, until the tops of the riffles are clear of gravel; then the water is shut off entirely.

The line boards are loosened and removed from the upper boxes, beginning at the lower end, after having been carefully washed off with a hose or a whiskbroom to remove any adhering gold. Then the riffles are taken out and cleaned in the same way.

Next, a strip of 1 in. by 2 in. lumber, the width of the box, is nailed temporarily across the lower end of the box, to act as a dam and catch any values that might escape when the water is turned on again. With a light flow of water, just sufficient to move the material, the concentrates are shovelled slowly and carefully up toward the upper end of the box with wooden paddles. These are made of 1-inch lumber, about 4 or 5 inches wide, tapering to a V at one end and shaped for a handle at the other, in length about 3 or 4 feet. The sluice box chosen for clean-up should have a good smooth bottom, free from knotholes or cracks.

In working over the concentrates against the current, the paddle is held flat side down about 45 degrees against the box bottom and pushed forward and back a foot or two, for the purpose of keeping the material well stirred up so that the gold will settle to the bottom, as in panning, while the black sands will be gradually carried off. Rocks and oversize material are picked out by hand and thrown away; the weight, if not the color, would reveal any nuggets. The gold and heavier sands gradually collect at the upper end; these concentrates are shovelled into buckets for further treatment elsewhere. By brushing the concentrates with a whiskbroom, while they are still in the box, some more black sand can be eliminated.

The concentrated material from the lower boxes, as many as are to be cleaned up, may be brought up to the first box in buckets and cleaned up at the same time, or the same process can be followed down the line.

Where there is only a single box to be cleaned up, it may save time to remove the riffles, and shovel all the concentrates into a bucket, replace riffles and set into sluicing again; the concentrates, if small in amount, can be conveniently taken care of later in a pan or rocker.

If desired, the crevices and cracks in the sluice box bottom can be examined for fine gold, which can be removed with a pointed instrument, which is sometimes amalgamated. After the final clean-up, the sluice box can be burned and ashes panned for values. As much as $20 has been removed from a single box in rich ground.

Treatment of bucket concentrates

The rich bucket concentrates are treated in a pan or rocker for a final product. If the gold is very fine and difficult to clean and save, a pan with a copper amalgam bottom can be used, and the amalgam subsequently taken off and retorted, or a small amount of quicksilver can be introduced into the concentrates, which will amalgamate the bright gold even in the finest sizes. The Chinese take the black sand and rub it by hand with the quicksilver in a receptacle, but this is not recommended. A better plan is to make up a dilute solution of potassium cyanide (*a deadly poison*) using a piece the size of a small marble to a gallon of water, and immerse the concentrates in it for 30 minutes or so. This will brighten up most rusty gold so as to make it amalgamable. Such cyanide solutions will *dissolve* fine gold if in contact too long. Instead of cyanide, a strong lye solution is said to have the same brightening effect on rusty gold.

If there is only a small amount of concentrates, they can be amalgamated in a strong fruit jar, with half their weight of water and quicksilver added. The correct amount of quicksilver to add depends on the amount of gold; it requires about 3 to 6 parts of mercury to 1 of gold to make a fluid amalgam. Hence, if it is estimated that the concentrates, 5 ounces in weight, are one-third gold, about 5 ounces or more of mercury are required. The jar is shaken for half an hour or so, when all the gold should be amalgamated. The amalgam is easily recovered by panning—the excess quicksilver not so easily. For further details on amalgamation, see Chapter IX. Use of a copper-amalgamated pan has already been suggested for separating gold from black sands in the final clean-up.

The Berdan pan

For amalgamating gold in the presence of black sands on any large scale, the Berdan pan used extensively in California by the old placer miners has much merit. It consists of a circular trough, 5 feet in diameter, set at 45 degrees off vertical, which is revolved at slow speed by gearing from a central shaft. A large steel ball weighing about 120 pounds is placed in the lower part of the trough and sits in a pool of mercury. As the trough revolves, the ball assists in grinding and amalgamating the gold. Water, introduced with the feed, overflows at the lip, carrying off the finer material. There is

Fig. 9.—Berdan pan.

practically no flouring of the mercury as the speed is low and the action quiet. Power needed is negligible, and the machine is not so messy as an amalgam barrel, which can also be used for the same purpose.

In field work, it must be clear that there never is any trouble about saving coarse gold, up to 10 mesh in size (about 1/16 of an inch diameter). It is the fine sizes that are lost. Not only the size, but the flattened, leaf-like shape of the fine particles makes them easy to lose. This is the only reason in sluicing for extending the string of sluice boxes beyond two or three, to give every possible opportunity for the fine gold to settle. The cost of additional boxes is so small that if an additional ounce or so of gold is recovered in the final box at the end of the season it probably has paid to put it in.

Undercurrents

Many ingenious schemes have been tried to save the fine flour gold and separate it from the black sands, and there are on the market scores of machines, more or less reputable, that claim to accomplish the purpose. A certain skepticism of such claims is not unwise. Many a machine will work well in the laboratory, with special sands and approved conditions, and fail utterly in the field.

Both theory and practice indicate that fine gold is best saved by spreading the gravel feed out in a broad, thin stream, increasing the grade, and cutting down the amount of water, with as much oversize material eliminated as possible. On this principle, undercurrents are built. An undercurrent in general is any device which spreads the water quickly over a large area, thus reducing its velocity and carrying capacity, and so permitting the fine gold to settle. There are several standard forms, one of which is shown in the sketch.

Construction of undercurrents

In construction, the bottom of one of the tail boxes is cut out in the center about 4 feet in length across the width, and steel bars 1 in. by 4 in. are set across slightly recessed, to form a grizzly, over which the coarse material will slide, while some of the water and most of the sands will drop through. The grizzly bars will be set fairly close, $\frac{1}{4}$ to $\frac{1}{2}$

inch apart, the proper interval being determined after experiment. If too much water comes through, it may cause the tail boxes in the main sluice to choke up. If the bars are set too close, they will clog up and let nothing pass. Instead of grizzly bars, a piece of heavy punched steel plate with ⅜-inch

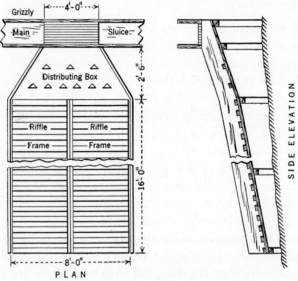

Fig. 10.—Undercurrent, plan and elevation.

holes can be used. An adjustable drop gate can be built beneath the grizzly to regulate the flow.

Below the grizzly or screen a launder somewhat wider than the sluice opening conducts the water and sands to the distributing box and tables. This launder is usually set at a fairly steep slope, at right angles to the main sluice. The distributing box has baffles to break up the water flow for even distribution. The tables are usually in two sections, in width 8 to 10 feet, with a much heavier grade than the main sluice, from 1 to 1½ inches per foot. The length may be 16 to 25 feet, with sides 8 inches high. The purpose of having two sections is to permit clean-up of one side while letting the other carry the entire flow.

Riffle arrangement on distributing tables

Riffles on the distributing tables may be made of ordinary wood strips, with sufficient spacing, ½ to 1 inch. An idea more in favor is to use cocoa matting, burlap, or corduroy laid flat on the bottom, and covered with ⅛-inch wire screening or expanded metal. Corduroy is excellent for saving fine gold values. It should be laid with the ridges across the water flow. In cleaning-up, the screening is taken off, the corduroy removed in a roll and rinsed with hose and whiskbroom in a tub of water to clean off the concentrates, which are subsequently panned. The corduroy is then replaced. The whole operation takes but a few minutes.

An undercurrent obviously assumes plenty of slope for the main sluice box, with a trestle support under which the launder to the undercurrent can be slung. When this condition is lacking, the so-called caribou riffle boxes can be tried for the same purpose, and many placer miners prefer them to the conventional undercurrent arrangement in any case, as the cost of installation is very much less.

Caribou riffle

The caribou riffle (**Fig. 11**) consists of perforated steel sheets placed over a blanket or corduroy in the tail box. These sheets are set in a series of low, flat steps, with the end of the sheet on the upstream side resting on the blanket, while the other is raised about 2 inches on wooden strips which support the full length of the sheet, coming to a point at the upper end. The arrangement is as shown in the sketch.

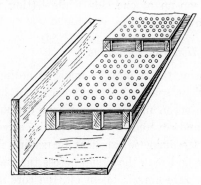

Fig. 11.—Caribou riffle, with one side of sluice box removed.

The sheets are ⅛- to ¼-inch metal, with ⅛-inch staggered holes, and are said to last about 2 months before needing replacement. Slot holes are not advised on account of clogging. If possible, the holes should be reamed out underneath

Fig. 12.—Arrangement for Caribou riffle and undercurrent.

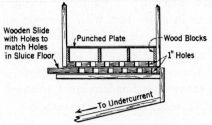

Wooden Slide with Holes to match Holes in Sluice Floor — Punched Plate — Wood Blocks — 1" Holes

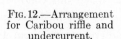
To Undercurrent

so as to give a clear passage. For cleaning-up, the screens and bridges are removed, and the corduroy washed as described under undercurrents. The slope of the tail box that has the caribou riffles will necessarily be steeper than the main sluice.

Figure 12 shows a section of a sluice box with a caribou riffle and undercurrent. The floor has been bored at the lower end with 1-inch holes, 3 inches apart, to take the place of a grizzly. A wooden slide with holes to match the holes in the floor regulates the discharge as desired by partly closing the openings. The fines drop through the openings, as the blanket has been cut to allow for the holes, and fall into the launder and so to the undercurrent.

Saving fine gold on Snake River

An interesting instance of saving fine gold is related by R. N. Bell* about the Snake River Basin, Idaho; it should have application elsewhere. Sluicing is done along the river bars exposed during low water. A favorable place is located along the first terraces, which rise 10 to 20 feet above high-water mark, and about 40 feet of 3- to 4-foot sluices are built, with a steel punched screen in the first box. A head of water is turned through the sluices, and the fine material going through the screen is diverted at right angles on both sides of the sluice box to distributing boxes, and thence to the under-current tables, 4 feet wide and 10 feet long. These are covered with canvas or burlap. The gold, though very fine, settles readily, and 90 per cent is caught on the first 4 feet of the canvas. It is swept into a trap every 4 hours by diverting the feed and turning on clear water. The concentrates, which appear to be coated with some oxide, are run into an arrastra, quicksilver added, and the gold amalgamated, after grinding and polishing in the mill. This is said to save 90 per cent of the values, although the gold is so fine as to require 1,500

* Eighth Annual Report, Idaho Mining Industry.

colors to make a cent. Obviously, ordinary methods of recovery would fail here.

A black sand trap

When placer deposits are exceptionally heavy in black sand, there is considerable trouble with the riffles packing. This requires more water, which carries off the fine gold that rides over the riffles. An ingenious black sand trap, described in Memoir 5 of the Montana Bureau of Mines, is shown in the

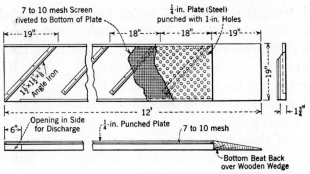

FIG. 13.—Black sand trap.

sketch. Unlike undercurrent devices, this trap is placed in the first sluice box, ahead of the first set of riffles, and the object is to get the black sand and the heavy gold out of the sluices at once for other treatment, giving the fine gold a chance to settle in the riffles.

The trap consists of a heavy galvanized-iron pan whose edges are turned up and crimped over, 1¾ inches high, along the two sides. The length recommended is 6 feet, with width the inside of the sluice box. Every 18 inches along its length 1½-inch angle irons are rivetted diagonally to the bottom.

On one side of the downstream end is a 6-inch opening for the flow of black sand and water into a tank. The upstream end is built into the shape of a wedge, and has the galvanized iron bent back over it to aid the movement of the gravel and water on to the plate and screen. These latter are built in 3-foot sections, of $\frac{1}{4}$-inch steel plate, with 1-inch punched holes, with a 7 to 10 mesh wire screen rivetted to the bottom. After the bottom or pan of the device has been placed in the sluice box, the plate and screen are laid on top. As the feed moves over the plates, the heavy black sands and gold fall into the trap and are carried out to a tank.

CHAPTER IX.

USE OF MERCURY IN PLACER MINING.
AMALGAMATION. RETORTING.

The use of mercury (quicksilver) in the pan, rocker, or sluice box, which is occasionally advocated as a cure-all for saving fine gold, is by no means always a profitable or economic procedure for the placer miner working on a small scale, though at times it is amply justified. It should not be used until simpler means for saving the values have been tried and found wanting.

In the first place, mercury is expensive and often difficult to obtain. It is ordinarily sold in 76-pound iron flasks, at a dollar a pound, though the market price fluctuates. In smaller quantities, it costs much more than this; drugstores asks $4 per pound. It is difficult to carry around; there is always a definite loss wherever it is used on the job, both in amalgamating and in retorting. Inexperienced operators will encounter trouble as well from "floured" and "sickened" mercury; and if they are careless about retorting, they stand a good chance to become salivated. It will not work effectively on rusty gold, and it is unnecessary to use it on coarse gold.

Using mercury in the sluice boxes

However, with bright, fine gold, that cannot be recovered without much time, labor, and loss, mercury introduced into the lower riffles in sufficient quantity yields an amalgam that is much more easily saved in the pan or rocker than the fine values. It is sometimes sprinkled from a cloth-covered bottle into the sluice, while the gravel is being run in, or it is poured into the riffles after clean-up. In any case, on account of its great weight (specific gravity, 13.6), it will lodge in the riffles. A few ounces of mercury per 12-foot box is enough, and a single flask will last a good size operation all season. The sluice box must be perfectly tight, and a mercury trap should be built at the tail box to catch any that splashes over and escapes. A fresh supply should be put in the sluice from time to time, depending on the condition of the riffles, amount of amalgam being recovered, etc. If an undercurrent is being used, mercury can be used to good advantage in a riffle below the blankets or corduroys.

Amalgamation—what takes place

When bright, clean gold of any size comes in contact with mercury, it is "wetted" by it. The gold undergoes penetration by the mercury, becomes brittle, and loses its color. The degree of penetration depends on the time of contact and the size of the gold particle. Hence there are amalgams of varying degrees of hardness and color. Coarse gold, that is little penetrated, yields rich amalgam; fine gold, poor amalgam. The amalgam, being heavier than the mercury, will sink to the bottom where there is an excess of the latter. Mercury will amalgamate with other metals besides gold, notably silver. In handling amalgams from clean-ups, always use iron or granite pails, not galvanized iron or tin, as the mercury will attack them.

Treatment of amalgam

Hard amalgam will not recover gold; it should be kept in a semi-liquid or plastic state by addition of more mercury. Amalgam as recovered from the riffles is generally dirty, and is contaminated with sand, iron, etc. After it is placed in an iron receptacle, additional mercury is added and stirred well until the whole is fluid and the impurities rise to the surface, when they are skimmed off. The amalgam and mercury are then poured through heavy, close-sewed canvas, or preferably chamois, and the free mercury filters through in drops which are caught in a vessel. The amalgam remaining behind in the chamois is squeezed dry by twisting the corners until it is a hard compact lump, when it is ready for retorting. Squeezed amalgam should contain about 1 part of gold to 2 parts mercury.

"Sickened" and "floured" mercury

Mercury may have various ailments after continued use. "Sickened" mercury, black in color, results from contact with base metals, oil or grease; hence these should never be used around it. In this condition it will not amalgamate the gold satisfactorily. By washing the "sickened" mercury with a weak solution of nitric acid, sodium cyanide, or lye, it may become active again. "Floured" mercury results from excess agitation or overgrinding. The mercury is broken up into a multitude of fine globules, white in color like flour. It is useless for amalgamating, and is easily lost in this condition. By pouring it back and forth into an excess of clean mercury it may be made to reunite.

Preparing an amalgamated plate

Although it is unlikely that the placer miner will have much need for amalgamated copper plates, a knowledge of how to prepare one may be useful. The copper plate, set on a smooth, flat surface, is first washed with a strong alkali solution to remove all grease and dirt, then scrubbed with Sapolio or a similar compound until bright. It is then cleaned with a dilute solution of cyanide, about 1 ounce to a gallon of water, and let dry overnight. Mercury is sprinkled on the surface and rubbed in until the hard, dry appearance is changed to a bright, moist one, when it is ready for use.

The plate can be used in the undercurrent, and is efficient for saving fine gold that comes in contact with it. Grade and amount of feed and water are determined by trial. Care should be taken that the amalgam does not get too hard; additional mercury should be added to soften it. The amalgam will not collect evenly all over the plate, but usually in ridges; if it gets too high, there is danger of scouring and loss. It is removed from the plate by using a paddle of hard rubber, or broad thin scraper of flexible steel.

Properties of mercury

Mercury starts to boil at 674 degrees F., about the same temperature that lead melts. As the mercury is volatilized, the gold remains behind as a sponge and is recovered. The gold is never completely freed from mercury even at a bright red heat. Retorted gold from clean amalgam should have a clear yellow color and be readily broken up from the sponge. If it is incompletely retorted, it will be light to dark gray on account of the mercury still present. Sulphur and arsenic will blacken and discolor sponge gold.

The mercury passes off as a grayish vapor, and is *extremely* poisonous to inhale. Salivation may result, with loss of teeth and general breakdown of health. With ordinary care and precaution, however, there is little danger.

Treating small amounts of amalgam

Small amounts of amalgam can be rolled in paper and placed over a forge on a shovel or iron spoon which has been coated with clay as a lute. A high heat is unnecessary to drive off the mercury. Of course, this method results in losing all the mercury, and unless a good draft is provided, there is danger from the fumes.

The potato method is used by prospectors for small balls of amalgam. A good-sized potato is cut in half, and a large enough hole scooped out to hold the amalgam. The two halves are then wired together and placed in a bed of hot coals. After the potato is baked, the gold is found in the cavity, and some of the mercury, which has condensed in the pulp, can be recovered and saved by crushing it. Or the amalgam can be put on a shovel, and half the potato, scooped out, used as a cover. Don't try to eat the potato afterward.

Another way of recovering the gold is to treat the amalgam with dilute nitric acid, which will dissolve the mercury and save the gold in fine needle-like crystals, which can be melted down in a crucible.

Retorting amalgam

For any considerable amounts of amalgam, however, a retort is required. This is not expensive: the ¼-pint size costs $4; 8-quart size, $12. A retort consists of a pot-shaped iron vessel, with a removable cover, which is held firmly in place by a yoke and screws. The gooseneck which conveys away the volatilized mercury should be large, with gradual bend, so there will be no choking.

A homemade retort can be constructed from pipe as shown in the sketch. Care should be taken that all joints are tight so that there will be no escaping mercury fumes. There may be some danger in the condensed mercury collecting in the bend in the pipe, which should be frequently tapped to dislodge it.*

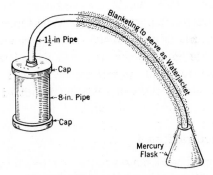

Blanketing to serve as Waterjacket

←1½-in Pipe

←Cap

←8-in. Pipe

←Cap

Mercury Flask

FIG. 14.—Home-made retort.

The retort must be thoroughly cleaned and coated on the inside with chalk or graphite to prevent "sticking." The amalgam, broken up and wrapped in paper, is placed inside the retort, never more than two-thirds full, and a little charcoal sprinkled on top. The rims of cover and retort are smeared with clay or wood ashes for luting, and the cover screwed on tight. If there is no water jacket over the delivery pipe from the gooseneck, the pipe should be covered with sacking, which should be constantly drenched with cold water to condense the mercury. Canvas should be wrapped around

* Suggested by J. B. Knaebel, Bureau of Mines.

the open end of the pipe to form a conduit for the mercury, and lead it into a pan of water. The level of the water should not be so high that water could be sucked back into the pipe and thence into the retort, in case the fire died down, as an explosion would result.

The retort is supported on a tripod, and a fire built around it. Heating should be gradual and will take about an hour or more at a dull, red heat. Too much heat will cause the re-tort to bulge out, owing to the weight of the amalgam. As volatilization goes on, the pipe should be tapped occasionally to make sure it is clear. When no more condensed mercury appears in the pan, the heat is raised for a few minutes to a bright cherry red, and the retort then allowed to cool gradu-ally. Be sure that the delivery pipe is out of the water at this time.

Refining retorted gold

The retorted gold may still be impure, though it can be sold to the Mint if practically free from contaminants. How-ever, it can be still further refined by placing it in a graphite crucible, with a flux of borax glass and a little soda, and heating to the melting point. Molten gold is bright green in color. For very small lots of bullion a clay crucible can be used, well coated on the inside with borax glass. After the gold is melted, the clay crucible is removed, cooled, and broken, when the gold can be separated from the slag on top by hammering. With graphite crucibles, the molten gold is poured into heated ingot moulds, with a little fine rosin sprinkled at the bottom. All crucibles, slags, and drippings, of course, should be saved for recovery of included gold.

Chapter X.

DRY PLACERS.

Dry placers, where water is entirely lacking and must be hauled in, or where the supply is so limited that it must be rigorously conserved, are obviously not adapted to the usual methods of placer mining. On account of their generally difficult living conditions, their inaccessibility, and to date the lack of success with most large-scale methods of handling them, dry placers have not received much attention, in spite of the fact that large arid areas in parts of Arizona, Nevada, and California are known to have considerable promise. In Australia, dry washing has accounted for a notable amount of the gold production.

Occurrence of gold in dry placers

The occurrence of gold in dry placers is somewhat different from that in more normal ground. As water action has been lacking, there is generally an absence of the rounded water-worn pebbles so characteristic of other gravel deposits. The gold, similarly, is often rough and angular, indicating a shorter travel from its source. Generally, though not always, the gravel is cemented together with iron oxide, or lime, and must be disintegrated and broken up before the gold can be recovered. The Mexican term for this material is "caliche." The values are scattered all through the cement, in size from fine specks to over a pennyweight, though nuggets of much larger size are not uncommon. The analogy of the occurrence of gold in the cement of desert placers with its occurrence in clay and fine sand on the side of boulders in creek placers is marked. Quite frequently rich pockets of gold are found in clay accumulations under large boulders, where the gold has been caught beneath the boulder in its sudden travel during a desert cloudburst. Such pockets are not necessarily on bedrock.

In the more arid regions, gold is sometimes found in the wash or alluvium, which has been carried downward by an occasional cloudburst in a torrent of water, sand, and gravel, without definite stratification. Naturally, such gold is extremely patchy in occurrence, is invariably coarse, and is mixed with the sand and gravel, imbedded in the clay, or clings to crevices in the bedrock if exposed. These wash deposits vary greatly in thickness. Sometimes the values are disseminated throughout the whole; or it may be necessary to sink shallow shafts to the pay-channel.

Some dry placers are covered with shifting sands on top, beneath which are jagged boulders, and finally just above the bedrock there is rich sand and gravel. In other cases the deposit is cemented from surface to bedrock. Greatest concentration of values usually occurs near where there has been the greatest erosion.

Conditions for working dry placers

Obviously, the value per yard of a dry placer must be substantially greater than of a creek or bar placer to make it workable under average conditions. Aside from the consideration of higher living costs, gold recovery from dry placers is usually much less than with the other, and is attended with higher costs. But with the prospect of leaner ground in the older placer districts, it is likely that the small-scale operator will give more attention in future to less favorably situated placers in the arid regions, where ingenuity, courage, and adoption of the right method of handling the ground economically may obviate the water difficulties.

In many cases a little water can be obtained either by pumping or building a small reservoir to conserve the water from infrequent showers. Although the amount obtained may be too little for sluicing or hydraulicking, it may permit rocking or operating a long-tom or a shaking table, by settling and re-using the water. For instance, the rocker can be set in a watertight box, with sides 10 inches high, and the water drained off from the box through plug holes below the surface as it gets clear. Similarly, the long-tom can be run with a short sluice, no attempt being made to save the very fine gold, and the water, caught in a sump, can be pumped back to the head end. Where water is entirely lacking, recourse must be had to dry blowing. Contrary to some of the common impressions, the dry blower, *if conditions are right,* can do remarkably good work, that on coarse gold (10 mesh or larger) will compare favorably with wet washing. The trouble is that ideal material is hard to find for dry washing. A little clay that cements the gold to the sand or gravel makes it impossible to use dry washers satisfactorily.

The Quenner crusher for cemented gravel

The primary conditions are that the material to be treated must be entirely dry, and well pulverized. Even a slight amount of dampness ruins the work of the dry blowing machine. Pulverization, with cemented gravel, may require machinery of some sort. Fortunately, pulverized gravel will often "air slack" and become amenable to treatment. To crush the entire mass in the ordinary rolls or crusher involves a lot of wasted energy, as there are no values within the pebbles or boulders, but only on the outside. The Quenner machine, supplied by Roy & Titcomb, Nogales, Arizona, has had some success for large-scale work with cemented conglomerate. This consists of a cylinder, the circumference of which is made of steel bars, through which passes a shaft. From the shaft, chains are attached in a spiral arrangement, and on its ends are manganese-steel hammers. These hammers, when extended by centrifugal force, clear the inside surface of the trommel by ¼ inch, and are suspended so as to strike edgewise. The shaft carrying the hammers revolves independently, and in an opposite direction, from the cylinder's revolution. The conglomerate is crushed into fragments, but not completely pulverized, and the worthless pebbles go out at one end, while the undersize material, passing through the spaces between the bars, is ready for the dry blowing machines. A 2-inch rim at the end of the cylinder catches and retains any heavy gold too large to pass through the holes.

The Mexican dry washer

All dry blowing machines are constructed along the same general lines. They consist essentially of a screen for taking out the oversize material, a hopper, a removable riffle board, and a bellows, all mounted on a wooden frame. A Mexican dry washer, which is used generally in the dry placers of the West, is shown in the cut in side elevation, from which the construction is clear. It is built of 2½ in. by ½ in. lumber, with screen 2 ft. by 1½ ft., and stands about 4 feet high. Before

Mexican dry washer. Courtesy of manufacturer,
J. O. H. Newby, Phoenix, Ariz.

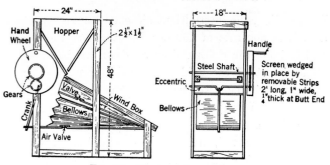

FIG. 15.—Mexican dry washer.
From Bulletin No. 8 University of Nevada.

operation, the machine should be carefully levelled up. Undersize material falls into the hopper, from which it flows to the
riffle tray, which is suspended over the windbox, with detachable wooden strips on the sides to keep the gravel from spill-
ing over and to permit easy removal for clean-up. The bottom
of the riffle tray is covered with coarse wire netting, over
which are stretched one or two layers of burlap and a layer of
thin cotton cloth, or canvas alone. Transverse riffles, $\frac{1}{2}$ inch
square, are spaced across the surface, 4 to 6 inches apart. The
riffle frame, constructed of 1 in. by $\frac{5}{8}$ in. strips, is 16 inches
wide and 38 inches long.

Underneath this tray is the windbox, which receives the
pulsations of air from the bellows. Two flap-valves permit in-
gress and egress of air as the bellows are worked. These valves
are located in the center of the windbox, and below it. The
bellows are opened and closed by a crank connected to an
eccentric from gears driven by the handwheel. The gear ratio
is such that one revolution of the handwheel will pump the
bellows three times.

Operation of a dry washer

In operation, the intermittent air puffs coming through the
cloth top at the rate of 150 per minute cause the sand and
gravel to jump the riffles and travel down the table and over
the end. Most of the gold and considerable black sand is
caught between the first two riffles. The dust and light particles are blown away.

After the machine has operated for some time, the side
boards are removed, and the material behind the riffles
brushed into gold pans. As this material is seldom completely
clean, it is fed back slowly once more into the hopper, and put

through again, when nearly all the gold remains behind the first riffle. The first two riffles are then brushed out clean and the concentrates are panned and the gold separated from the black sand.

Capacity and performances

The capacity of this type of dry washer varies widely, from $1\frac{1}{2}$ to as much as 12 cubic yards per day, average about 3 yards when operated with two men, depending on the nature of the material. J. V. Richards* gives the capacity at Sonora, Mexico, where it was operated by Indians, as 2 to $2\frac{1}{2}$ cubic yards per hour, but this seems too high. The same writer comments on the excellent recovery of gold made by the machine, which according to his tests was 87.5 per cent of the coarse gold in the gravel. However, making allowance for the loss of fine gold by dusting he estimated 80 per cent overall recovery—a creditable figure. An instance reported by Geo. A. Packard† at Round Mountain, Nevada, gives 17 tons (about 12 yards) of gravel put through in 10 hours with two men, working a gravel bench 1 to 6 feet in height with a recovery of 70 per cent.

Some other interesting details are given. Large rocks were picked out of the gravel, which was thrown against a screen with 1-inch openings, and then shovelled into dry washing machines, which had a top screen with $\frac{1}{4}$-inch openings. About 100 shovelfuls were thrown on by one man while the machine was being cranked by another. This took about 15 minutes. The machine was then stopped, the screen removed and replaced with another. Concentrates were dropped in a

* Transactions, A. I. M. E., Vol. 41, pp. 797-802.
† Engineering & Mining Journal, Vol. 83, pp. 150-151.

tub, and later re-run over the machine. They were finally cleaned by panning. The tails from the second run were saved for shipment and were reported to have $40 per ton in gold.

Difficulties in dry washing

The canvas covering of the riffle frame must be replaced after running through about 150 yards of gravel. Cost of building a dry blower of the type illustrated is only about $25 to $40. Aside from the inconvenience of working continually in a cloud of fine dust, the chief difficulty is the uncertainty of the weather, for the least dampness will cause any clay in the gravel to "ball up" and ruin any recovery of the gold. Also, if most of the gold values are in the very finest sizes, recovery is similarly very difficult. But for fairly coarse gold, it has been demonstrated that this simple machine will do good work. Large nuggets, of course, if present, will not go through the screen; hence before discarding the oversize it is a good plan to look for such rarities in the gravel. When the gravel is cemented and must be pulverized by hand, much of the operator's time is taken up with preparing the gravel for treatment—a tedious and laborious process. If there is nothing more than a hand mortar for the job, 800 to 1,000 pounds of feed is all that can be expected, hence value per yard must be very high to justify it.

There have been several moderately successful efforts to work dry placers on a large scale by a combination of dry and wet methods, but any application of air separation for large-scale work has to date been unsuccessful.

CHAPTER XI.

PLACER MINING MACHINES.

Small scale placer mining is hard work. There is always much physical labor in panning, rocking, or sluicing, and as is always the case when muscle is employed instead of mechanical power for heavy jobs, the cost of operation is high. Consequently, the gravel must contain a relatively large amount of gold to make hand methods of work profitable. On the other side of the picture are healthful working conditions in the open air, generally low living costs, and the ever-present lure of finding if not a bonanza, at least some unexpected rich ground that will bring up the general average.

With the exhaustion of the richer ground it becomes necessary to work those areas where the gravels are leaner, and this means not only that attention must be given to recovering the fine gold, but that a much greater yardage of ground must be worked per day than was possible by the old methods. The problem of the placer miner today is just as much or more the economical mining, transportation, and disposal of the gravel as it is the saving of the gold within it. Native ingenuity in getting the gravel cheaply to the sluice box has made many a lean piece of ground pay out.

In designing a small placer machine for recovering gold from the generally low grade ground now open to prospecting, the manufacturers therefore had two primary objects in mind: to increase the amount of gravel that could be treated in a

See Author's Note, page 145.

day's time over what could be done by hand methods, and to save the very fine particles of gold that formerly escaped recovery. This they have generally succeeded in doing, although they have had more success with the first than the second objective. In addition, they had to make the machine portable, or semi-portable—not so heavy as to make it impossible without undue expense to transport from one location to another. It had to be strongly built, without parts that would break down from hard work in the field, far from a machine shop. It had to be economical of water for use in arid regions, or for summer work when the streams were nearly dry. Finally, it had to work without skilled supervision, it must operate at a low cost, and be sold at a low price.

There are said to be over 7000 gold saving devices patented, most of which exist only in the laboratory of some well-meaning but inexperienced enthusiast. More than 150 individual machines are described briefly in a recent publication of the California State Department of Mines.* Some of these are in actual operation in the field but the great majority have little but the word of the manufacturer or inventor to substantiate their claims and results of actual field work are lacking on which to base judgment. In this connection, it must be admitted that it is difficult to get much exact information from users. Most of the operators are not technically trained and have little advance knowledge of what their ground will run and consequently cannot tell whether they are making good recoveries. Too often a machine is blamed and a manufacturer condemned for failure to recover the gold when the plain truth is that sufficient values were never in the ground originally to justify the purchase of a machine.

* Bulletin, California State Division of Mines, July, 1934.

There are some things that no placer machine can do. It cannot take 50 cents worth of gold out of gravel that carries only 25 cents, yet too often enthusiasm may make one believe that it can accomplish the impossible. It is doubtful if any machine or mechanical gold-saving device can make a higher recovery than can be obtained by careful, painstaking hand work, though it may do it with a minimum of the labor and in much less time.

In general the successful manufacturers of placer machines since 1930 have given less attention to the extremely small light machines that were originally designed, and made succeeding models heavier and more efficient mechanically. Ball bearings have been introduced and protected from slime and grit when in operation. Riffles of mechanical design have been replaced with rubber mats that are easily replaceable and can be quickly removed for cleanup. The problem of handling clayey gravels has been recognized, as well as the need for thorough disintegration to prevent clay from coating the fine gold particles with subsequent difficulty in amalgamating and danger of their being carried over with the tails in heavy slimy water. This has led in some instances to the installation of a scrubber and trommel screen ahead of the gold saving devices.

Before buying any placer machine, it should be tried out beforehand on a generous sample of the ground upon which it is expected to work, to determine actual, not theoretical recoveries. The expense of sending on several hundred pounds of gravel, more if possible, for testing is fully justified. Reputable manufacturers welcome such opportunities for preliminary trial. "The best criterion of a novel piece of apparatus is in

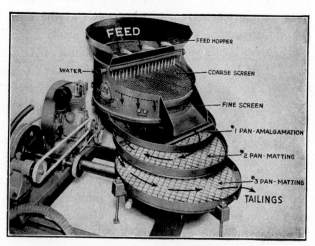

Above: Denver Mechanical Gold Pan in operation.
Below: Construction of Denver Mechanical Gold Pan.

experience gained from working it under field conditions on a sufficiently large scale for a period long enough to truly demonstrate what a machine will do."*

And finally, the old recipe for making squirrel pie is worth remembering—"First catch your squirrel." Be sure that you have a piece of ground that from careful testing and examination will repay the capital cost of whatever type of equipment is wished to put in.

Descriptions of a few of the better-known placer machines are given in the following pages. These machines have demonstrated their merit in the field by actual test since the gold rush of 1930 started and their manufacturers have been able since then to make improvements in design from hard facts of experience. The list of course is limited; a complete category would require a larger book than this to enumerate, and information on lesser-known machines can be obtained from the California Bulletin referred to. Some of the larger installations used in recovering placer gold are briefly mentioned for those interested in more extensive operations.

The Denver Mechanical Gold Pan

The Denver Mechanical Gold Pan, manufactured by the Denver Equipment Company, Denver, Colorado, is one of the best known of the placer machines and has found use not only in the western states but in placer fields in other parts of the world. The appearance and construction of the machine are plainly shown in the cuts. In its operation it is designed to imitate the hand-panning motion by means of a steel yoke and saddle connected to an enclosed rotating ball-bearing eccentric

* Placer Mining in Nevada, A. M. Smith and W. O. Vanderberg.

with the nest of pans tilted slightly forward. A positive vibrating motion of 240 oscillations per minute results.

When the machine is operated without a trommel attachment, gravel is shoveled directly into the feed hopper where it is washed with water from the spray pipe and works down over the upper screen, made of $\frac{1}{4}$-inch heavy punched plate. The oversize is thrown out as tailings, while the undersize drops to a fine screen below and the minus 1/16-inch material falls through the latter on the first of the oscillating pans. These are two feet in diameter.

The top pan has a copper-amalgamated bottom for amalgamating immediately all free gold possible. The overflow lip on the edge of the pan causes a bed of sand $\frac{1}{2}$ inch deep to form on the pan and the pulsating motion permits a settling action which allows the gold to settle to the bottom and amalgamate.

The sands then flow by gravity to the second pan made of steel with rubber matting at the bottom over which is fastened a 1-inch screen to hold the latter in place, as well as to act as a riffle to catch the fine gold and any dislodged amalgam. The third pan is similar in construction to the second, and is used to insure recovery of any remaining values that may have escaped the first and second pans. Thus the screened material has three separate treatments before being discarded. All the pans are interchangeable and the plates and matting can be removed quickly for cleanup.

Power is furnished by a $\frac{3}{4}$-horsepower gasoline engine, which also drives a $\frac{3}{4}$-inch centrifugal pump for supplying water. A larger size engine is recommended for altitudes above 6500 feet, and according to some users a 1-1$\frac{1}{2}$-horsepower engine is generally more suitable although it means more weight. Gasoline and oil requirements are low, about 1 gallon of gasoline for 8-10 hours run, and a pint of oil is enough for 30 hours operation. The cost of fuel should not exceed 3-4 cents per yard.

Emphasis is placed by the manufacturers on all metal and welded construction throughout, with bearings protected from contact with sands and slimes. The machine weighs about 400 pounds net, shipping weight 650 pounds. List price, with engine, pump, and 10 feet of suction hose, $195.

The machine is claimed to use a minimum amount of water for operation, and if water be reclaimed, it can be reduced to one part of water to one of gravel put through, that is, about 300-400 gallons per hour. The ¾-inch pump, running full, will provide about 25 gallons per min. Naturally, the amount of water required will vary with the type of gravel being treated, the amount of clay in the gravel, and the rate of feed.

The machine has a rated capacity of 1½-2 yards of bank run gravel per hour, but so much depends on the character of gravel and amount of clay that average operating capacity will probably be less than that. With loose, clean gravel as much as 3 yards per hour are reported; with clayey gravel the capacity is much less, as considerable time is lost in getting the gravel to disintegrate properly. A mudbox may be required where the feed is particularly clayey. For larger operations with clayey gravels the manufacturers of the machine offer a special trommel attachment with screen which breaks up the material first in a closed section of the trommel, then gives it a scalping treatment in the revolving screen and discharges the oversize. Of course this is recommended only for semi-permanent installations.

Good recovery of both coarse and fine gold is reported by

Above: Model 34 G-B Placer Machine showing flexible molded
rubber riffle.
Below: Model 34 G-B Placer Machine in operation.

users of this machine. While it is difficult to get any exact figures from the field, recoveries of 80-90 per cent of the gold are not uncommon. Difficulty has been experienced in saving gold when clay is unusually heavy in the gravel, as this makes amalgamation difficult as well as making it harder to cause the flour gold to settle. Much better results can always be obtained when the feed is regular and steady.

The machine has been used successfully by engineers for testing work in large placer areas as it handles much more ground than is possible with hand methods in a day's time. It is also used for making cleanups from sluicing operations, and is reported to have reduced 2500 pounds of heavy black sand concentrates from the underflow of a sluice to small bulk in an hour's time, without loss of gold.

The G-B Placer Machine

This machine was originally introduced in 1931 by the Mine and Smelter Supply Company of Denver, Colorado. A large number of them have been sold to placer mine operators in this country and abroad for testing purposes and small scale work in alluvials. The new Model 34 differs in many respects from the machine as first placed on the market, but the general principle of operation remains unchanged.

In the original model the gravel was shoveled into the upper launder from which it was washed to a grizzly screen which screened out and discharged the oversize. It was found in practical operation that while this arrangement was satisfactory for gravels that were uncemented and free from clay, it did not work out so well when clay was present. There was little chance for the material to become thoroughly disintegrated in its passage down the launder, and there was likelihood of the clayey material washing over the grizzly and carrying away fine gold.

The machine was therefore redesigned with the addition of a combination scrubber and screen, which receives the feed from the hopper and subjects it to a thorough washing and disintegrating action before passing the fine material to the undersize launder and riffle. This change in design necessarily increases the weight of the machine considerably, necessitating heavier construction throughout, and made it suitable for semi-permanent installation.

The appearance of the Model 34 G-B machine is shown in the cuts illustrating the machine in operation from which the design is readily apparent.

In operation the tanks, of 90 gallons capacity (about 3 barrels) are first filled with water and the 1-horsepower gasoline engine started. Gravel is shoveled into the feed hopper, where it is sprayed with water as it passes into the revolving scrubber and screen. The fine material passes into the outer jacket and thence directly to the undersize launder and on to the riffle. The oversize receives a final washing and further rolling and tumbling before being discharged through the chute. The scrubber is encircled with a depressed space at the discharge end for catching any coarse nuggets.

A flexible molded rubber riffle is used in the riffle pan of the Model 34 machine, which represents a change from the original design that operated without this accessory. The riffle operates crosswise at an average speed of 200 strokes per minute, receiving motion from an eccentric shaft. Riffle corrugations are so shaped, it is claimed, as to hold the gold while the gravel moves on to the discharge end of the pan. The riffle pan is readily removable when required; how often depends on the amount of gold and black sand in the feed. The washed sand drops into the conveyor tank where it is dewatered and discharged by the tailings conveyor flights.

The machine is economical of water and therefore has special application where it must be conserved. By means of the tank arrangement a partial clarification is effected, and unless there is an unusual amount of slimes, it can be used over again. The separation of the external settling tank into two compartments permits most of the settled slimes to be shoveled out. Make-up water, that is water to be added to take care of the amount absorbed in wetting the gravel, is about 60-75 gallons per yard of gravel treated. This of course is a minimum figure and assumes that the tank water would be used indefinitely.

The weight of the Model 34 machine, uncrated, is about 600 pounds and crated for shipment, 825 pounds. It can be taken apart for muleback transportation or loaded on a trailer without difficulty. Obviously, however, it can hardly be called a "portable" machine, in the sense that it could be readily moved from place to place where roads were lacking. List price, packed for shipment, is $335 FOB Denver.

The machine has a rated capacity of 2 cubic yards bank run gravel per hour when the amount of clay or cemented material is not excessive, but reports from users show somewhat less capacity than this under field conditions. So much depends on varying factors that individual figures do not mean much.

Cost of operation is nominal. The gasoline engine requires about a gallon of fuel for 8 hours run. Recovery of gold is reported as 85-95 per cent, depending, of course, on the amount of flour gold present. There is apparently no difficulty in recovering the larger particles. It is reported that in South

Africa where the machine had been used in saprolitic material, averaging about 50-75 cents per yard, that a close recovery has been obtained in the fine sizes. An operator in a western state, working on channel gravel, reports 95 per cent recovery from $3 gravel, with the machine handling 10 yards in 8 hours, at a cost of 3 cents per yard. With clayey ground he reports a 75 per cent recovery. This gravel, it should be noted, is far above the average in value.

A large number of machines based on the principle of using some application of centrifugal force to effect a separation of the gold and black sands from the gravel have been patented. The Ainley Centrifugal Gold Separator, manufactured by the company of the same name at Denver, is the best known of this group. In this machine separation is made in a revolving bowl mounted under a revolving screen. Gravel is fed through a hopper into the screen, and the fine material passes through the screen with water into the bowl. Centrifugal force holds the heavy gold and black sands in the riffles of the bowl while the waste material passes over the top. The cleanup is made through a plug valve in the bottom of the bowl. It is claimed that only 15 gallons of water per minute are required, and the water can be used over again, as muddy water does not affect recovery.

The machine is made in several sizes. The smallest size, Model B, a 12-inch bowl, is recommended by the manufacturers for preliminary test work in placer ground preparatory to installing larger units if results are satisfactory. Capacity is given as $\frac{1}{2}$ to $1\frac{1}{2}$ cubic yards of material that have passed through a 5-mesh screen. The bowl operates at 285-300 R.P.M. and requires 1/3 horsepower. Gravel must be thoroughly disintegrated before being fed into the machine. Weight of the small bowl is 180 pounds, with gears, drive shaft, pulleys, tailings and feed spout mounted on a steel frame.

A larger size, 36-inch bowl, is sold in a set of four bowls for large scale work, adopted for power shovel or drag line feed with proper screening facilities to eliminate material coarser than ⅜-inch. Capacity of each 36-inch bowl is given as 10-15 cubic yards per hour of bank run gravel, or 5-10 yards of minus ⅜-inch material. The large size bowl employs a removable rubber riffle instead of having the riffles machined into the inner surface of the bowl as was formerly done. It is claimed these rubber riffles outlast the iron ones and can be replaced without removing the bowl from its frame.

According to a western operator, a battery of four 36-inch bowls using a 1¼-yard power shovel and customary screening equipment is treating 50-60 yards of sandy and rocky gravel with some clay and gravel cemented with the lime, making a recovery of 98 per cent of the gold. The gravel carries 50-60 cents per yard. Water requirements are 300 gallons per minute. Cost of operation is about 20 cents per yard. No difficulty is reported in separating the gold from the black sands, and the cleanup is shipped direct to the Mint.

The W.S.C. Placer Mill

This machine is not manufactured by any company, but was designed by Dean A. E. Drucker assisted by G. E. Ingersoll of the School of Mines and Geology, Washington State College, with the idea of producing a portable plant at small cost that could be constructed by any placer miner with mechanical ingenuity who was able to obtain the necessary materials. It was designed to recover fine flake and flour gold from river gravels, and is reported to have operated successfully on low grade gravels from the Columbia River. Its construction involves several interesting points, in particular the use of a shaking table, somewhat similar to a vibrating undercurrent, instead of the devices used for gold saving described in the machines previously mentioned. The machine is described in

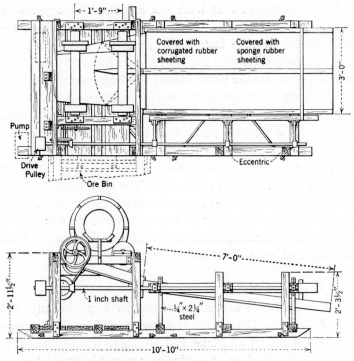

Plan and Elevation of W.S.C. Placer Mill.

detail in Information Circular No. 2 issued by the College. The cut shows plan and elevation.

Essentially, the machine consists of an orebin receiving

screened feed, an ore-feeding device to the trommel, and a shaking table covered with rubber riffles for recovering the gold. The theory involved is that the horizontal vibration of the table will keep the black sands in suspension while permitting the gold to settle out and be held by the rubber riffles. The slope of the table can be varied and the amount of water changed depending on the feed in order to get the best results.

In operation the gravel is screened over two large coarse screens covering the orebin before entering the trommel. This bin has a sloping bottom and an automatic feeding device actuated by cam and springs which causes the feeder to shake and permits the feed to enter the trommel at a uniform rate. The trommel has $\frac{1}{4}$ inch holes and rests on trunnions. It is driven by chain drive to one of the rollers from a jack shaft, and slopes 1-inch to the foot. It is bound on the outside with 14 mesh wire screen to provide a 2-inch clearance outside the drum. Angle irons bolted inside the upper part of the drum serve to break up the gravel and disintegrate the clay. Oversize from the trommel is discharged as waste.

A sloping slide or distributor head below the trommel has an adjustable gate, so pivoted as to swing to either side and divert the feed to left or right or to both sides of the table as may be desired.

The unique feature of the machine is the shaking table for recovering the gold. This is 7 feet long and 3 feet wide, with a strip running down the middle which makes it really a double table. By diverting feed to one side, clean-up or adjustment can be made on the other while in operation. The upper 6 feet of the table is covered with corrugated rubber sheeting, while the last foot is covered with sponge rubber with a strip of corrugated iron at the bottom in which mercury can be placed.

The table is supported by four steel uprights. Its slope can be adjusted by means of holes in the uprights at the lower end, so that it can be raised or lowered to get the best results. The vibrating motion is obtained by three eccentrics on the line shaft which give a ¼ inch movement to the table. Three connecting rods transmit this movement to a horizontal timber which in turn transmits the movement to the table. The table normally vibrates 400 times a minute; this can be varied by changing the size of drive pulley on the line shaft.

To cleanup, the gate is swung over to cover the side desired, and clean water is run over the riffles while vibrating to remove surplus material. The rubber sheeting is removed and washed in a tub of water. When sufficient concentrates have accumulated, they can be treated in a cleanup barrel with mercury and the gold amalgamated. It is suggested that the concentrates be first ground with some pebbles to scour off the oxide film of the rusty gold, if present, before adding mercury.

Power is furnished by a 1½ horse-power gasoline engine, connected by belt drive to a pulley on the line shaft. Another belt drives a 1-inch centrifugal pump which provides water through a perforated pipe to the trommel and thence to the shaking table.

The capacity of the machine is estimated at 1½-3 cubic yards of placer gravel per hour depending on the material treated. Tests have shown a recovery of 90 per cent and upwards of fine gold passing 14 mesh.

Bendelari Diaphragm Jig-Bulolo Type

For larger scale work where it is proposed to handle upwards of 20 yards per hour, the Bulolo type jig, a development of the Bendelari Diaphragm Jig used successfully in the lead and zinc areas of the Tri-State field, has demonstrated its ability to save very fine gold. A description of its performance at Bulolo Gold Dredging Ltd., appears in Engineering and Mining Journal, May, 1934.

The installation at this property consisted of three 42-inch jigs in series. However, a two-cell unit is recommended for placer gold feed except where there are values in the very fine black sands, when three cells are needed. Smaller cells than 42-inch are not recommended by the manufacturer as they cost too much in relation to their capacity. Some doubt is expressed if such cells would save all the fine gold as the travel of the feed is shorter, but it is believed that they would be more efficient than tables particularly with coated gold.

Capacity for either the two- or three-cell unit of the 42-inch jig is given as 30-40 yards per hour when the feed is 60 per cent plus ¼-inch, 40 per cent minus ¼-inch and the latter never exceeds this limit in the initial feed. The feed should be disintegrated ahead of the jig, but slimy water will not prevent effective concentration of the gold in these machines. Best results are obtained with placer feed when there is a flood of water over the jigs, that is, all the water from the trommel and up to 75 gallons per minute per cell for pulsion water.

Portable large-scale washing plants

Several large-scale portable washing plants have been built in the West operating on gravels with a value of not less than 35 cents per cubic yard. The gravel is mined and delivered with a stationary drag scraper or a caterpillar-mounted shovel or dragline. The feed hopper is placed either above the screen or at the base, with apron feeder or bucket elevator respectively to regulate the feed. A wash pipe sprays water on the gravel as it enters the feed hopper while another pipe with spray nozzles placed inside the full length of the screen helps to disintegrate the gravel. Several types of gold saving equipment are employed, including Ainley bowls, shaking amalgamators, jigs, or sluices. Tails are stacked with a belt conveyor. Make-up water is about 50 gallons per minute for a 1000-yard daily capacity. The cost is upwards of $5,000, outside of excavator. In most cases the plant is built to order to suit special operating conditions at the property where it is to be installed.

Stacker scow and excavator for dredging ground

Another plan to work dredging ground of limited area that will not justify the cost of a large dredge is to install a stacker scow to float in a pond in the dredgable material, which serves as a water reservoir as well. The ground is excavated with a drag-line excavator, dumped into a hopper on the scow where it passes through a rotating screen, gold saving tables, and sluices with riffles. Tailings are stacked in the rear of the scow by a conveyor. The machinery is mounted on a scow consisting of three steel pontoons. Such an installation is adapted for dredging ground where depth of gravel does not exceed 20 feet and other conditions such as grade, bedrock, water, etc., are suitable. Definite advantages exist where yardage of profitable gravel is sufficient to justify initial cost. With a 1½-cubic yard excavator, capacity of such a plant is about 1000 yards a day. Cost, exclusive of excavator is upwards of $12,-000. As built by the New York Engineering Co., New York, such a plant handling about 20,000 cubic yards per month has shown an operating cost of 17 cents per yard. Under favorable conditions, operating cost should not exceed 12 cents per cubic yard.

Chapter XII

LOCATION OF PLACER MINING CLAIMS. LEASING SALE OF PLACER GOLD. SELLING A PLACER PROPERTY.

For the old-time prospector, it was just as necessary to know the procedure in locating a claim as to discover some valuable ground. Every new area that had any promise was followed by a rush to stake all open ground at once, and it was necessary to act promptly to protect one's rights. At the present time, much of the promising placer ground in the western states is either owned outright or held under the mining law by performance of annual assessment work, and no outsider can work such ground without making suitable arrangements with the owner, usually by leasing and paying a royalty on the gold recovered, as there is no provision under the laws of the United States for locating mineral claims upon or within privately owned lands.

However, there are still large areas open in Montana, Idaho, Arizona, Nevada, and parts of California, that fall under the domain of public lands, which can be located as claims, and the prospector should know about the United States statutes, as well as the state and local regulations, in order to protect himself.

Before entering placer fields in state or foreign country, a copy of the revised mining codes should be obtained from the proper state authority, if it is thought that some ground may be located. There are certain minor points in procedure that differ in the various states. If prospecting in foreign countries, the United States Bureau of Mines should be consulted for individual reports on the mining laws of the country in question. Sixty-one such reports are available.

Public lands open for location

According to the Revised Statutes of the United States, all valuable mineral deposits in lands belonging to the United States, both surveyed and unsurveyed, are free and open to exploration and purchase by citizens of this country. Where lands have been previously surveyed, the entry shall conform to the legal subdivisions of the public lands. Legal subdivisions of 40 acres may be subdivided into 10-acre tracts.

Each placer claim, if located by one person or a corporation, may not exceed 20 acres in area, and should measure 660 ft. by 1,320 ft. The smallest tract recognized is 10 acres and measures 660 feet. If a claim is located on unsurveyed land, it should have the same dimensions and shape as if it were on surveyed land, and the boundaries must run north and south, east and west. Where this is impracticable, as in the case of gulch placers, the lines should run as nearly as possible in these directions.

The law permits two persons to locate a 40-acre placer claim, three persons 60 acres, etc., but the largest allowable claim is 160 acres for eight persons. An individual or a corporation may locate any number of placer claims.

Locating a claim

A discovery of mineral is necessary for each location of a placer claim. The grade of the gravel does not have to be profitable, but it must be good enough to encourage the doing of development work on the claim.

After discovery of mineral, these steps must be taken: *First:* Post a location notice. Printed forms are generally used, but any paper will serve that gives the name of the claim, the name of the locator or locators, date of location, number of acres claimed, and a description of the claim, by such reference to natural objects or permanent monuments as will identify it. This notice must be posted conspicuously at the point of discovery. It is a good plan to put duplicate location notices on the corners of the claim as well. *Second:* Mark the corners of the claim at each angle by a wooden post or monument. The kind of monuments that are legal varies in different states, as well as the time allowed for erecting them. *Third:* Within a period varying from 30 to 90 days, depending upon the state, the claim must be recorded with the county recorder of the county in which the claim is situated by a location certificate. This will give the same information posted on the location notice. *Fourth:* To retain title to placer ground that has been legally located, improvements or work amounting to $100 must be done on each claim between July 1 following the date of location of the claim, and the following July 1. This annual assessment work must be done each succeeding year until a patent is obtained. The character of this work is specified in the statutes of the respective states, as well as details of making affidavits as to its accomplishment.

With regard to the United States surveys of public lands, it is usual for one or two sections of land in each township to be reserved to the state as "school lands." Mining rights within these "school sections" cannot be initiated under the United States mining laws, but must be acquired from the state, usually by lease. The prospector should therefore ascertain what sections within any township are reserved as "school lands."

If a piece of placer ground turns out to be so valuable as to justify patenting, a lawyer should be consulted for the necessary steps.

Leasing

Wherever placer ground is privately owned, a lease or other equitable arrangement should be drawn up with the owner before starting work. Although it is true that objection is seldom made to very small-scale work, such as panning or rocking, it is always better to be on the safe side, so that if the ground develops unexpectedly well, there will be no danger of the lessee losing his rights to mine. Fair-minded men can draw up a workable lease that will give adequate protection to each party without legal assistance. If large-scale work is contemplated, however, with such problems as diverting water supply, disposal of tailings, road construction, erection of a camp, etc., it is best to get legal advice.

Royalty is usually set at 10 to 15 per cent of the net mint return. Sometimes a sliding scale of royalties can be arranged depending on the grade of the gravel as determined from regular clean-ups. Profit-sharing schemes are generally unsatisfactory.

It is generally necessary to specify in the lease that a certain minimum amount of work must be done during the life of the lease. Damage done to the surface of the land by tailings, excavations, etc., must be considered in advance, as well as all questions on obtaining adequate water supply. If more than one party is working on the ground, the rights of each one must be protected in the lease.

Special inquiry should be made in regard to water rights, which may have been pre-empted for irrigation or other purposes; where clayey gravels are being washed, water may become contaminated for stock. No plans for extended placer work should be made until these are looked into.

Sale of placer gold

The United States assay offices, at the most convenient points, are immediate and satisfactory purchasers of placer gold, in amounts as low as 2 troy ounces. From Form 506, United States Mint Service, the following information is obtained.

Gold is received as bars, also as lumps, grains, and dust in their natural state, or nearly so. Sponge gold (the product after retorting off the mercury from amalgam) is also received, but raw amalgam is not accepted. Gold may be brought to the mint in person, or it can be sent by mail or express, charges prepaid. A letter must be sent separately, giving instructions, with description, approximate weight, estimated value, as well as locality in which it was obtained. Payments are made 2 to 5 days after shipment is received.

Payment is made for all the gold and silver in a shipment, but no allowance is made for any platinum that may be included. Charges of $1 are made for melting, of $4 for a bullion assay, to determine the purity of the gold, and a separating and refining charge of ½ to 8½ cents per ounce. Gold is paid for at $34.912 per ounce; silver at 64½ cents (December, 1934).

Obviously, the larger the amount of gold sent in a shipment, the smaller the ratio that the charges bear to the value of the shipment. For example, if 10 ounces are sent in, the statement might appear:

Weight of shipment ....10.0 ounces
Weight of bullion after melting..................... 9.4 ounces
Fineness of bullion as determined by bullion assay .... .825
Calculation; 9.4 ounces @ 0.825   7.795 fine ounces
Value at $34.91 = $272.12
Less refining charges, app.   5.25
                                        ―――――――
            Net...........$266.87

Selling a placer property

The placer miner may find a piece of ground that in his opinion has good possibilities for a large company with ample capital, but which he is totally unable to work on a small scale. It will be suitable, presumably, for dredging or hydraulicking.

Enthusiasm without the facts will not find a purchaser for him. The more data he can collect and present in convincing form, the better chance he will have of interesting a large operator to send out an engineer to investigate and check up. All the points suggested in Chapter II should receive attention, with such confirming figures as he can get.

The results of his actual test work are especially valuable, and should be as complete as possible, with all tests, good and bad, included.

A map of the ground, of course, is needed, with topographic details. This need not be elaborate, but must be approximately correct. A pacing survey, with pocket compass and hand level for elevations, will serve, if done with care. Photographs add much, and often make a crude map intelligible.

Samples of the gravel will not mean much; a log of the test pits will carry more conviction. Samples of the gold recovered in clean-up should certainly be shown; their appearance, size, fineness, etc., mean a great deal to the engineer, and there is a psychology in viewing the actual raw metal that always results in immediate attention and favorable interest.

Often all that the vendor can hope to do at first is to persuade his prospect to examine the property, either personally or by his engineer, at the latter's expense. Of course, it is assumed that the property has been definitely tied up with an option or sales agreement so that the vendor is adequately protected, and all negotiations must go through his hands. It is generally too much to expect that a sale can be consummated, or even a substantial payment obtained, without preliminary check-up and examination. Willingness to co-operate with the examination, and, if suggested, to accompany the examining engineer, are favorably regarded by interested parties.

Such points as maintaining a good appearance, having self-confidence, and avoiding any impression of being in need of money, or requiring an immediate large advance payment, require no emphasis. This is just good selling psychology, the world over. Personal approach is preferable to selling through an agent, and to place a property with an obscure or ill-informed broker generally results in disappointment. Most mining engineers, if not interested themselves, will often suggest names of others who would investigate possibilities of meritorious properties.

BIBLIOGRAPHY

The following books are standard authorities on the subjects they treat, and are recommended to those who want more detailed information on placer mining, as well as to all who wish to work on a large scale.

WILSON, E. B., Hydraulic and Placer Mining, 3d Edition, John Wiley & Sons, Inc., New York, 1918, 425 pp. Besides conventional placer mining methods and equipment, has valuable data on hydraulicking, calculation of water flow, pipe-line installation, dredging, etc.

BOWIE, A. J., JR., Practical Treatise on Hydraulic Mining, 11th Edition, D. Van Nostrand Co., New York, 1910, 313 pp. Especially good on California practice.

PEELE, R., Mining Engineer's Handbook, John Wiley & Sons, Inc., New York. The section of this standard reference book devoted to alluvial mining presents an immense amount of data on the whole subject in small compass.

TAGGART, ARTHUR, Handbook of Ore Dressing, John Wiley & Sons, Inc., New York. Special sections devoted to recovery of placer gold, riffle construction, clean-up, etc.

YOUNG, GEO. J., Elements of Mining, McGraw-Hill Book Co., New York. Chapter on alluvial mining treats the whole subject comprehensively.

LONGRIDGE, C. C., Hydraulic Mining, Mining Journal, London, 1910, 352 pp. One of the older authorities, well worth consulting for reference on practice outside of the United States.

VON BERNEWITZ, M. W., Handbook for Prospectors, McGraw-Hill Book Co., New York, 359 pp. A good general handbook on the whole subject. Considerable space given to gold.

IDRIESS, I. L., Prospecting for Gold, Angus & Robertson, Sydney, Australia, 272 pp. Highly interesting and practically written. Describes Australian practice and operating methods for both placers and lode mining.

VOLL, MAX, A. B. C. of Practical Placer Mining, Great Western Publishing Co., Denver, 62 pp. Written by an experienced placer miner. Contains useful information for the man with small capital.

GOVERNMENT PUBLICATIONS

Both the Geological Survey and the Bureau of Mines have contributed valuable publications on special subjects.

PURINGTON, C. W., Methods and Costs of Gravel and Placer Mining in Alaska, U. S. Geol. Survey, Bulletin 263 (1905), 272 pp., illustrated with many photographs from the field, as well as cuts of equipment and mining methods. Prospecting, water supply, hydraulicking, open-cut methods, costs, and a wealth of general operating data in Alaska.

WIMLER, N. L., Placer Mining and Costs in Alaska. Bureau of Mines, Bulletin 259, 1927, 236 pp. Brings Purington's paper up to date, with much new material, especially on dredging. Both these publications can be studied with profit by placer miners everywhere.

GARDNER, E. D., and JOHNSON, C. H., Placer Mining in Western United States. Part 1. General Information, Hand-Shoveling, and Ground Sluicing. U. S. Bureau of Mines Information Circular 6786, 1934, 77 pp., 9 figures. A comprehensive treatise with considerable historical data on placer gold production in U. S. Good material on costs of testpitting, drilling, shoveling-in and ground sluicing. Equipment and provisions for prospecting outfits. Subsequent papers will describe hydraulicking and dredging. Strongly recommended.

STATE PUBLICATIONS

ARIZONA. Arizona Gold Placers and Placering, by E. D. Wilson and others, University of Arizona, Tucson, 1932, 132 pp. Maps and descriptions of all Arizona districts where placer gold has been found. Small-scale placer mining methods and equipment. Location of placer claims. Equipment for desert prospecting.

CALIFORNIA. Gold Placers of California—Mining Methods, by C. S. Haley, California State Mining Bureau, Bulletin 92, 1923, 167 pp. A standard reference book, fully illustrated, that should be in every placer miner's library.
Mining in California, April, 1932, Vol. 28, No. 2 of the Reports of the State Mineralogist, 113 pp. Chiefly interesting for description (without comment) of 117 machines and processes for wet and dry placer mining.

IDAHO. Elementary Methods of Placer Mining, by W. W. Staley, Idaho Bureau of Mines and Geology, Moscow, Pamphlet 35, 1932, 23 pp. Brief description of placer areas in Idaho, and description of simple placer equipment.

MONTANA. Placer-Mining Possibilities in Montana, by O. A. Dingman. Montana School of Mines, Butte, Memoir 5, 1932, 33 pp. Brief description of placer districts of Montana, with comments on production and possibilities. Placer Mining equipment described and well illustrated.

NEVADA. Placer Mining in Nevada, by W. O. Vanderburg and A. M. Smith, State Bureau of Mines, Reno, 1932, 104 pp. A complete and well-prepared account of placer operations in the state, with many excellent photographs and operating data. Especially valuable for account of dry placer methods on moderate scale.

BRITISH COLUMBIA. Placer Mining in British Columbia, by J. D. Galloway, Department of Mines, Victoria, 1931, 107 pp. Descriptive and illustrative.

Besides the foregoing, the technical press has printed dozens of informative articles on various phases of placer work. The list is too long to be given here. By referring to the various indices of technical literature, the articles of special interest can be consulted at any large library. Papers of special importance on placer work have appeared in Engineering and Mining Journal, New York; Mining Magazine, London; Mining Journal, Phoenix, Arizona; and the publications of the American Institute of Mining and Metallurgical Engineers, including Mining and Metallurgy. Back files of the Mining and Scientific Press, San Francisco (no longer published) contain a great deal of helpful material, especially on hydraulicking. Late developments in dredge construction and operating are adequately covered in the technical press generally.

APPENDIX

TABLES AND CONVERSION DATA

Troy Weights Used in Weighing Gold

24 grains = 1 pennyweight
20 pennyweights = 1 Troy ounce = 480 grains
12 ounces = 1 pound

A troy ounce is *not* the same as an avoirdupois ounce, nor is a troy pound the same as an avoirdupois pound.

To convert avoirdupois ounces into troy ounces, multiply by 0.911. In other words, a troy ounce is about 10 per cent heavier than an avoirdupois ounce.

Metric Equivalents

1 kilogram = 1000 grams = 2.2 pounds avoirdupois
1 gram = 1000 milligrams = 15.43 grains
1 milligram = 0.015 grain
1 liter = 2.1 pints = 0.26 gallon
A liter is approximately 1 quart.

Water Equivalents

4 quarts = 1 gallon = 231 cubic inches
31½ gallons = 1 barrel
Weight of 1 gallon of water = 8.34 pounds
Weight of 1 cubic foot of water = 62.4 pounds
1 cubic foot of water contains 7.5 gallons, approximately
1 miner's inch = 1½ cubic feet per minute
= 11.25 gallons per minute

Value of Gold

	Old statutory basis	*New price*
1 ounce, 1000 fine	$20.67	$35.00
1 pennyweight	1.033	1.75
1 grain	4.3 cents	7.3 cents
1 gram	66.3 cents	1.126
1 milligram	0.066 cent	0.126 cent

137

Value of Gold per Troy Ounce at Different Degrees of Fineness

Fineness	Value at $20.67	Value at $35
700	$14.47	$24.50
800	$16.53	$28.00
900	$18.60	$31.50
1000	$20.67	$35.00

Specific Gravity of Common Minerals and Rocks (Water = 1.0)

Quartz	2.6
Mica	2.7
Ilmenite (black sand)	4.5
Rutile (black sand)	4.1
Magnetite (black sand)	5.1
Pyrite	5.0
Galena	7.4
Silver	10-11
Gold	15.6-19.4
Platinum	14-19

Tyler Standard Screen Scale

Mesh	Opening in millimeters	Nearest fraction of an inch
3	6.68	¼
4	4.70	3/16
6	3.32	⅛
8	2.36	3/32
10	1.65	1/16
14	1.16	3/64
20	0.83	1/32
35	0.41	1/64

Duty of Small Giants, Nozzle Diam. 1 inch

Showing gallons discharged per minute, and horizontal and vertical distances in feet reached by jets*

Pressure in pounds	20	30	40	50	60	70	80	90	100
Head in feet	46.2	69.3	92.4	115.5	138.6	161.7	184.8	207.9	231.0
Gallons per minute	110	134	155	173	189	205	219	232	245
Horizontal distance	70	90	109	126	142	156	168	178	186
Vertical distance	43	62	79	94	108	121	131	140	148

* From a table published by Mining and Scientific Press, 1904, p. 226.

Comparison of Hydraulic Mining Operating Costs*

Mines using water from	Number of mines	Cubic yards of gravel excavated daily		Tons excavated per labor shift	Operating cost per ton
		Mininum	Maximum		
Natural head......	26	40	500	41.25	0.042
Pumps	3	18	500	56.0	0.328

Data on 1-inch Belt-Driven Centrifugal Pumps
(from Fairbanks Morse Co.)

Size of discharge	Head	R.P.M.	Belt H.P.	Capacity, gallons per minute
1 inch ....20 ft.		1450	0.63	25
1 inch ....20 ft.		1770	0.90	50
1 inch ....20 ft.		2300	1.70	75
1 inch ....30 ft.		1750	0.73	25
1 inch ....30 ft.		2070	1.10	50
1 inch ....30 ft.		2420	1.85	75

* From "Sand and Gravel Excavation," Part 2. U.S.B.M. Inf. Cir. 6814, by J. R. Thoenen, E. M.

INDEX

A

Ainley centrifugal gold separator, 120
Amalgam retorting, 98
Amalgam treatment, 96, 98
Amalgamated plate, preparing, 97
Amalgamation, 30, 95
  in fruit jar, 85
  in pan, 29

B

Batea, 26
Bedrock, crevices, 23
  prospecting to, 22
Bendelari Jig, 124
Berdan pan, 85
Bibliography, governmental and state publications, 134
  text books, 134
Black sand, in placers, 16, 23, 33
  separation from gold, 29, 30, 45, 83, 85, 92
Black sand trap, 92
Blowing-down, 77
Booming, 71
Boulders in placers, 13
Breckenridge placers, 4
Bucket concentrates, method of treating, 84

C

Caliche, 101
Canvas hose hydraulicking, 75

Centrifugal pump data, 139
Clay in placer ground, 16, 46, 51, 54, 55
Cleaning-up, 82
  sluice boxes, 83
Cocoa matting, 66, 89
Colors, estimating value of, 33
Concentration of gold in placers, 9
Corduroy linings, 66, 89
Crevicing, 23
Cyanide for aiding amalgamation, 85

D

Denver mechanical gold pan, 113
Dredging ground, 20
Drilling placer ground, 14
Dry blower (or washer), 104
Dry placers, conditions for working, 102
  conserving water in, 103
  gold in, 101

E

Estimating value, of colors, 33
  of placer ground, 12
Estimating water flow, 50

F

Fire assays, 16
Flood gold, 10
Floor linings, 66

G

G-B placer mining machine, 117
Geology, of dry placers, 102
  of saprolites, 2
  of true placers, 3
Gold, associated minerals with, 32
  classification by screen sizes, 34
  examination of grains, 7
  fineness, 17, 132
  flour, 35
  occurrence, in dry placers, 101
    in stream placers, 8
  pan, 25
  properties of, 31
  refining, 100
  sale of, 131
  value of, 137
Gravel, character of, 15
Ground sluicing, 69
  with team, 70

H

Hydraulicking, 74
  canvas hose in, 75
Hydraulic mining costs, 139

K

Knock-down rocker, 41

L

Leasing, 124
Location of placer claims, 127
Long-tom, 59
  operation, 61
  slope, 61

M

Magnet, 28
Mercury, 94, 97
  "sickened" and "floured," 96
  use in panning, 29
  use in rocking, 45
  use in sluice boxes, 94
Metric equivalents, 137
Mexican dry washer, 104
  at Round Mountain, 107
  at Sonora, 107
  capacity and performance, 107
  operation, 106
Miners' inch, 51
  duty of, 53
Mint charges, 131
Mudbox, 46, 55

N

Nicaraguan placer, 2

P

Pan, copper amalgamated, 29
  gold, 25
Panning, 25
  operation, 27
  yardage per day, 30
Pay-streak, position of, 8
Placer claims, locating, 127
Placer ground, examination ot, 12
  in Arizona, 81
  in Nevada, 79
  methods of working, 78
  selling, 126
Placer mining machines, 109
Placers, bar, 6
  beach, 6

Placers—(*Continued*)
  bench, 6
  buried, 6
  dry, 101
  gulch, 10
  residual, 2
  resorted, 23
  stream, 4, 7, 10
Portable washing plants, 125
Prospecting, equipment, 21
  favorable places, 8, 9
  hints, 20, 22

Q

Quenner crusher, 103
Quick silver, *see* Mercury

R

Refining gold, 100
Retort, home made, 99
Retorting, 99
Riffles, block, 64
  caribou, 89
  construction, 63
  cross, 65
  Hungarian, 65
  in rocker, 40
  in undercurrent, 89
  longitudinal, 63
  number needed, 68
  pole, 62
  rubber, 66
Rocker, apron, 39
  cleaning riffles, 44
  construction, 36
  hopper box, 39
  knock-down, 41
  operation, 43

Rocker—(*Continued*)
  steel, 46
  water needed, 45
Rusty gold, 17

S

Sale of placer gold, 131
Sampling, 13
Sand in placers, 23
Saprolites, 2
Screen scale, 138
Shovelling-in, 72
  yardage handled, 73
Size of gold particles, 34
Sluice boxes, butt-end, 58
  construction, 57
  cost, 59
  grade, 51
  how laid, 54
  number needed, 54
  telescopic, 58
  velocity of water in, 53
Sluice head, 53
Sluicing, amount of water needed, 50
  field instances, 53, 56
  general discussion, 49
Snake River, 91
Specific gravity table, 138
Stacker scow, 126

T

Tables and conversion data, 137
Tailings disposal, 55
Test pits, 13

Transporting gravel, 73
Trough washer, 26
Troy weights, 137

U

Undercurrents, 87
  and caribou riffles, 90
  at Snake River, 91
  grade of, 89
  tables with, 89

W

Water carrying power, 52
Water equivalents, 137
Water in placer formation, 4
Water supply, for hydraulicking,
  75, 138
  for placer mining machines, 115,
  118
  for rocker, 45
  for sluicing, 51, 53, 70
W.S.C. Placer Mill, 121

AUTHOR'S NOTE

Since this book was originally published some of the placer mining equipment described in Chapter 11 is no longer being manufactured. The Mine and Smelter Supply Co. advise us that they discontinued manufacture of their G B Portable Placer Machine during the war years and have no plans to resume its manufacture.

On the other hand, Denver Equipment Co. is still manufacturing and selling its Mechanical Gold Pan, which is essentially as described in Chapter 11. The company informs us that over 700 of these machines are used by operators all over the world. Numerous improvements have been made during the last 25 years. The machine can now be bought in several sizes, to treat from 1 cu. yd. to 4 to 6 cu. yds. per hour. Prices vary from $820 for the smaller machine to $3215 for the largest.

The same company manufactures a trommel jig placer unit for recovering gold and other heavy minerals. Capacity of the smaller machine is 1–2 cu. yds. per hour, and of the larger, 4–6 cu. yds. per hour. Price of the unit, equipped with trommel, water pump, suction hose, and 4 H.P. gasoline engine ranges from $2850 to $3470, depending on size. Complete details can be obtained from the manufacturer.

The California Division of Mines, Ferry Bldg., San Francisco, notes that a number of small prices of equipment are advertised from time to time in the California Mining Journal. Attention is directed to an excellent pamphlet published by the Mineral Information Service of the Division of Mines, entitled "Skin Diving For Gold In California," describing operation and the equipment needed.

August 30, 1960